CIVIL SOCIETY AND STATE RELATIONS IN SWEDEN

Civil Society and State Relations in Sweden

MICHELE MICHELETTI
Department of Political Science
University of Stockholm
S–106 91 Stockholm
Sweden
tel. 046 8 16 26 31

Routledge
Taylor & Francis Group
LONDON AND NEW YORK

First published in paperback 2024
First published in hardback 2020

First published 1995 by Ashgate Publishing

Published 2024 by Routledge
4 Park Square, Milton Park, Abingdon, Oxon OX14 4RN

and by Routledge
605 Third Avenue, New York, NY 10158

Routledge is an imprint of the Taylor & Francis Group, an informa business

Publisher's Note
The publisher has gone to great lengths to ensure the quality of this reprint but points out that some imperfections in the original copies may be apparent.

A CIP catalogue record for this book is available from the British Library.

Library of Congress Catalog Card Number: 94-73717

ISBN: 978-1-85972-037-0 (hbk)
ISBN: 978-1-138-26753-4 (pbk)
ISBN: 978-1-315-26013-6 (ebk)

DOI: 10.4324/9781315260136

Contents

Figures

Appendices

Abbreviations

ABF	Workers' Educational Association (founded in 1912)
AMSA	Ad Hoc Group Against Swedish Atomic Bomb (founded in 1958)
ATP	Earnings-related (supplementary) pension plan
BIFO	NGO Foreign Aid Information Center (founded in 1992)
CND	Campaign for Nuclear Disarmament
DACO	Salaried Employees' Central Organization (founded 1931)
DFFG	United FNL Groups (anti-Vietnam war movement organization) (founded in 1967)
Ds	Reports from Cabinet Ministries
EFS	Evangelical Fatherland Foundation (founded in 1856)
ETUC	European Trade Union Confederation (founded in 1973)
FBF	Fredrika-Bremer-Association (founded in 1884)
Fi	Ministry of Finance
FNL	Vietnamese Front for National Liberation

FR	Swedish Free Church Council
FS	Non-Conformist Churches Study Association (founded in 1947)
HSB	HSB Association (national housing cooperative) (founded in 1924)
IF	Swedish Federation of Industry (founded in 1910)
IGO	International Government Organization
IKFF	Swedish branch of Women's International League for Peace and Freedom (founded 1919)
INGO	International Non-Governmental Organization
IOGT	Swedish Grand Lodge of the International Order of Good Templars (founded in 1879)
IOGT/NTO	Swedish Grand Lodge of the International Order of Good Templars/National Order of Templars (merger 1970)
KDS	Christian Democratic Party
KF	Cooperative Union and Wholesale Society (founded in 1899)
KK	Women's Consumer Guild Union, later called Consumer Guilds Union (founded in 1907)
LKPR	National Association for Women's Suffrage
LO	Swedish Trade Union Confederation (founded in 1898)
LRF	Federation of Swedish Farmers (founded in 1971)
MBL	Co-Determination Act
MHF	Abstaining Motorists' Association of Sweden (founded in 1926)
MIGRI	National Association of Environmental Groups, later called Environmental Guardians (founded in 1971)
MP	Member of Parliament

MRR	Swedish Civil Rights' Movement (founded in 1974)
NBV	Study Association of Teetotalers (merger of the IOGT and NTO study associations in 1971)
NIMBY	Not In My BackYard
NVO	Workers' Temperance Society, more commonly known as Verdandi (founded in 1896)
'old TCO'	Central Organization of Salaried Workers (founded in 1937)
PHM	Campaign Against Economic Planning
PRO	National Association of Pensioners (founded in 1942)
RFHL	Swedish Association for Aid to Drug Abusers (founded in 1965)
RFSL	Swedish Federation for Gay and Lesbian Rights (founded in 1950)
RGSM	Reference Group for Social Movements in government ministry
RLF	National Farmers' Union (founded in 1929)
RPG	National Association for Pensioner Solidarity (founded in 1974)
SAC	Central Organization of Swedish Workers (founded in 1910)
SACO	Swedish Confederation of Professional Associations (founded in 1947)
SACO/SR	Swedish Confederation of Professional Associations/ Federation of Government Officials (merger 1975), later called SACO
SAF	Swedish Employers' Association (founded in 1902)
SAFAD	Swedish Agency for Administrative Development
SAL	Swedish Agricultural Society (founded in 1917)
SALF	Swedish Union of Supervisors and Foremen (founded in 1905)

SAP	Swedish Social Democratic Party (founded in 1889)
SCB	Statistics Sweden
SCC	Swedish Co-operative Center (founded in 1958)
SFSF	Swedish Peace and Arbitration Association (founded in 1883)
SIF	Swedish Union of Clerical and Technical Employees in Industry (founded in 1920)
SKAF	Swedish Municipal Workers' Union (founded in 1936)
SKMF	Swedish Alliance for Women for Suffrage and Equal Citizenship (founded in 1921)
SKfV	Swedish Committee for Vietnam
SL	National Federation of Agricultural Cooperatives replaced the SAL)
SLR	Swedish Farmers' Supply and Crop Marketing Association founded in 1905)
SNF	Swedish Conservation Society (founded in 1909)
SOU	Reports from Government Commissions
SPR	Swedish Pensioners' Association (founded in 1939)
SPRF	National Association of State Pensioners (founded in 1937)
SR	Federation of Government Officials (founded in 1919)
SRK	Swedish Red Cross (founded in 1865)
TCO	Central Organization of Salaried Workers, later called Central Organization of Professional Workers (founded in 1944)
UD	Ministry of Foreign Affairs

Introduction

Sweden, as many other prosperous nations, is presently reassessing its national goals, political culture, and collective identity. Newer groups in society are demanding equal treatment, and others, whose struggles for recognition are older and unwon, are successfully mobilizing support for political change. Social democratic political hegemony has eroded, and other and new political forces are now reinterpreting past political achievements. Swedish history in being rewritten. There is a search for new ideas and methods of action and a need for historical perspective. Social scientists play a crucial role in times of reassessment. Use of theory and comparison put current demands for change in perspective.

This book analyzes the history of Swedish civil society. Social movements and interest organizations have played crucial roles in Sweden. Their history is also Swedish history and concerns struggles for political recognition and welfare state development and cutbacks. Theoretical developments within sociology, social psychology, public choice, and political science are combined to enrich the analysis. Some of the theoretical elements used in this book are organizational waves of development, organizational life cycles, political opportunity structure, and typologies of collective action organizations. Chapter 1 summarizes the theoretical approaches developed for this study.

The book deals with Swedish civil society history from the mid-1800s to the 1990s. The role played by collective action organizations in the important developments in politics, society, and economy in this one hundred and fifty year period are described, compared, and analyzed. The primary focus is the impact of change brought about by these developments on collective action organizations.

Swedish civil society history is divided into six periods. Period one is called formation of modern Swedish civil society and is dated from the 1850s to the 1890s. Many important collective action organizations emerged and began to place demands on the political order. They struggled for such political rights as freedom of religion and assembly, and some of them fought for economic justice. The second period covers the early 1900s (1900-1920s) and concerns Sweden's democratization and social movement institutionalization. Several of the social movements created in the first period mobilized sufficient support to force the political order to acknowledge them as legitimate political actors in the second period. Sweden became a political democracy, and a new political opportunity structure began to take form. All adults were given the right to vote; political parties began to develop; proportional representation was introduced, and Sweden conducted its first national referendum. These events had positive and negative effects on collective action organizations. The teetotaling, free church, women's, and peace movements began to lose public support, while others as labor unions and adult education associations flourished. Countervailing powers emerged to counteract the influence of established collective action organizations. The struggle for economic rights that began in the first period continued, as witnessed by the emergence of housing and farmers' cooperatives and growing support for the consumer cooperative movement. The second period of Swedish civil society history was one of the most conflictual in the twentieth century.

Civil society and government began to merge into the 'strong society' in the third period dated to the 1930s. Swedish politics became class politics, but at the same time this period was much less conflictual than the second one. A foreign scholar captured this nicely in the characteristic of Swedish politics as 'the middle way'. National identity and unity was stressed. Labor market organizations and new collective action organizations emerged; citizen empowerment was a main goal. The 1940s and 1950s form the fourth period of civil society history. The period is called the Swedish model. The welfare state developed; government expanded its operations, and the social democrats were able to gain a hegemonic hold on Swedish politics. There was collaboration across social classes, but as pointed out in chapter 5 the relationship between the two main labor market counterparts was more characteristic of a cold war than true class cooperation and compromise. Welfare state expansion impacted several collective action organizations negatively, as many of their tasks were taken over by government. Public responsibility-taking and corporatism characterize the relationship between collective action organizations and government. In the 1950s, several established organizations began to experience problems, and certain others began to stagnate.

The decisionmaking style of the Swedish model was questioned from the late 1950s to the 1990s. In the fourth period is called rethinking politics and the dissatisfied generation (1960s-1970s). Challenges to the political system were prevalent. New left and right forces confronted social democratic hegemony. Sweden's reputation for labor market peace and the middle way was tarnished. New political actors emerged and rejected the underpinnings of the Swedish model. New political issues challenged class-based politics. Postmaterial values became important. This period has been called a struggle for personal integrity and against all kinds of destruction. Clash over the nuclear power issue forced one government to resign. Women mobilized for political recognition. This was how Sweden experienced the 1960s. Changes in the political opportunity structure led to government instability. The viability of corporatism was questioned, and setting a boundary between civil society and the expanding state was discussed. The idea of contracting out welfare state services to collective action organizations became a political agenda item.

The impact of many of these developments can be seen in the sixth period, which is entitled repluralization and resurgence of Swedish civil society. It is dated to the 1980s and 1990s. The political system is undergoing transition, and this is affecting collective action organizations. New political parties have entered parliament, and social democratic hegemony has ended. New lines of conflict overshadow left-right political concerns. Gender is a top political priority. Organized business decided to withdraw from the corporatist exchange, and corporatism has been found to be a less desirable method for political policymaking. Labor unions are highly impacted by these developments. They are experiencing internal problems as well, and renewal is of utmost importance. Welfare state problems have created a need for a volunteering sector; several smaller voluntary associations have emerged to help solve social problems. Contracting out was emphasized by the non-socialist coalition government, which came to power in 1991. In fall 1994, the social democrats returned to power with an agenda that emphasizes balancing the state budget and lowering unemployment.

A thematic analysis of the six periods is offered in chapter 8. There are interesting similarities between the developments in the first two periods of Swedish civil society history and those of the last two periods. These blocks of time represent ages of transition with corresponding struggles for recognition. It is argued that political actors of the ages wanted to sign a new social contract. The first social contract involved such political rights as universal suffrage and freedom of religion, speech, and assembly. The second one concerns more conscious application of these rights to ensure that all individuals are treated equally. Swedish historians have characterized the first age by the terms internationalization, democratization, urbanization, individualization, and industrialization. Swedish political scientists view the second age as a time of internationalization, individualization, postindustrialization, decentralization and marketization. These characteristics are expanded upon in the following chapters.

With this book, I have completed my research ambitions regarding interest organizations and social movements in Sweden. This book is an abridged version of the Swedish volume *Det civila samhället och staten– medborgarsammanslutningarnas roll i svensk politik* (Stockholm: Publica, 1994). Most Swedish references have been deleted from this English edition. Several people have helped me write this book. The Bank of Sweden Tercentenary Fond has generously supported this project to employ assistants, attend conferences, purchase books and office equipment, and pay my salary. Max Edling entered the project at a late date. He did most of the research for the section entitled 'Granting Legitimacy to Collective Action Organizations' in chapter 8 and helped tie many of the loose ends together in other chapters. Fredrik Kron wrote a report on the SFSF. Johan Hall studied the topic LO and wage earner funds and wrote a paper on the Swedish model. Lorenz Jolin-Williams wrote a report on government grants to collective action organizations. Ann Närling was given the challenging task of investigating the opinion formation activities, strategies, and tactics of several collective action organizations. She interviewed their media experts and read organizational documents. Her systematic analysis has helped structure the discussion of the topic in this book. Tobias Wikström studied a number of smaller organizations concerned with social problems. He interviewed organizations spokespersons, and his well-written report has helped me draw general conclusions on the activities and situations of Swedish collective action organizations. Kristina Ahlén read some of the book and offered valuable comments. Nils Elvander read most of the manuscript. Olof Petersson cheered me up and showed me how to use two computer programs. Seven years ago, Claes Linde, our principle of undergraduate education at the time at the Department of Political Science, University of Stockholm, asked me to develop a course on interest organizations and Swedish politics. This was a golden opportunity to try out ideas on students, and assistants were recruited from it. The course was given for several years for Swedish students and American students, who were enrolled in the Swedish Program at the University of Stockholm. It was later replaced by a less comprehensive course on civil society and state relations in Sweden. I want to thank all students who took the course for their comments and encouragements and all guest speakers for coming to the course to inform us about their organizations. The assistance offered by the staff at the University of Stockholm Library and Royal Library as well as officials and staffs of the numerous collective action organizations that have been contacted in the process of researching this book has been appreciated greatly.

1 Analyzing change in Swedish civil society

A main task for social science is analysis of the major changes that have occurred in politics in the twentieth century. This task is especially important today as people seek a better understanding of contemporary problems by reflecting upon their historical roots. An important part of the Swedish political heritage is social movements and interest organizations. This chapter provides an analytical perspective for studying Swedish collective action over time. It summarizes the theoretical approach developed by the author in a companion volume, *Understanding organized action*. The reader may choose to refer to this book for references to scholarly writings not cited here to save space. The term 'collective action organization' is used to refer to people who work together for a goal and, thus, engage in collective action. The groups studied include ad hoc networks, social movements and interest organizations.

Civil society and government

Perhaps the concept of civil society is best understood when it is contrasted with the concept of mass society. *Mass society* is the naked society, a society without collective action organizations that clothe and protect individuals from excesses of government. Organizations of civil society play a crucial role in balancing tensions between individuals and government. Individuals come together in ad hoc groups, social movements, political parties, and interest organizations to act collectively on matters that concern them. Along with private enterprise and the mass media, these organizations constitute civil society. *Civil society* is a filter between the people and government. The main characteristic of organizations of civil society is their independence from government. The importance of collective action organizations for democratic political development is underscored in the role they play to help rethink progress, redefine democracy, and change the boundaries of the political.

Civil societies are well-developed in all democratic nations and absent in totalitarian ones. All collective action organizations do not promote freedom and democracy. The history of the national socialist movement in Germany shows that the actions of collective action organizations can promote political repression rather than democracy. Anti-democratic tendencies that emerge in civil society and gain public support impact government politics and institutions.

The important point in this and other discussions on civil society is that there must be a balance between political order, which is the responsibility of government, and individual freedom, which is guaranteed by a well-functioning civil society. The balance between order and freedom is never given, cannot be decided for future generations, and must constantly be reconsidered. Public debate on this balance occurs in times of great social, political, and economic transition.

This book shows how collective action organizations have safeguarded and furthered the development towards justice, democracy, and equality in Sweden. They have been instrumental in reevaluating the role that government plays in Sweden by questioning accepted distinctions of private and public responsibility. Figure 1.1 illustrates how they have broadened our conception of political democracy by struggling with government to enhance the number of issues included as inherent citizen rights. The characterizations of the political sphere and struggles included in the figure are explained more fully in the following chapters as well as in the companion volume.

On several occasions, collective action organizations have defended the political order in Sweden and other countries. They behave in this fashion when their supporters believe that their rights have been or can be recognized by government and when they feel that their interests are threatened by other collective action organizations which struggle with government and demand recognition of newer and different rights. Such developments show that collective action organizations evolve over time and that their position on the political order changes after they become established, legitimate political actors. Position on the political order is determined by the kind of social change sought by collective action organizations. There are four basic categories here: reactionary; conservative; revisionary, and revolutionary.

Collective action organizations that advance aims that are no longer held by the general public are *reactionary*. Supporters of such organizations want a return to the past. Frequently this reaction involves romantic notions of the good old days and a belief that older values are better than new ones. *Conservative* movements seek to maintain the status quo by exerting effort to

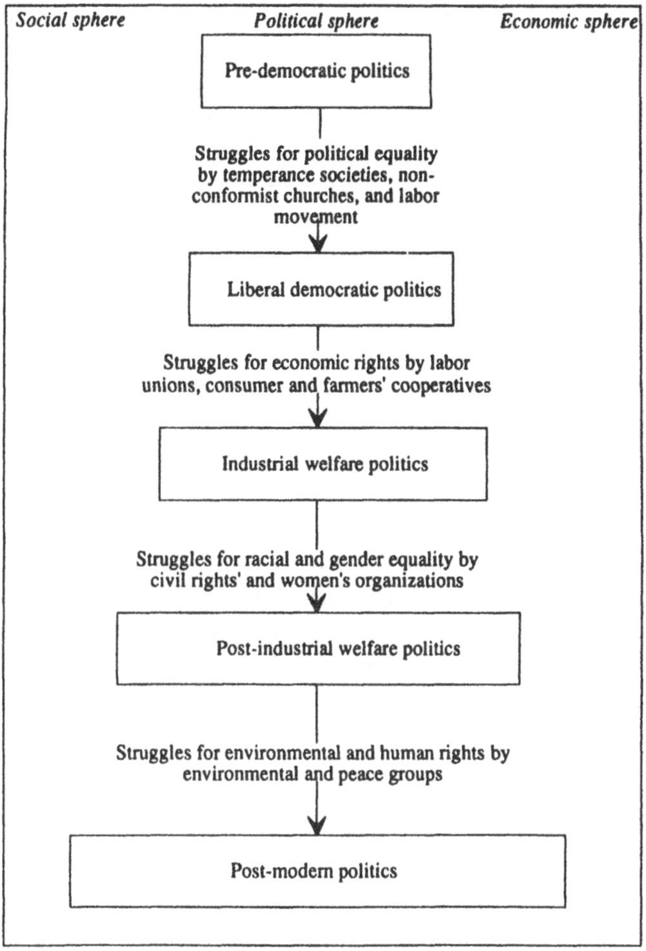

Figure 1.1 Role of collective action organizations in redefining democracy

obstruct change. They are created when change is forthcoming. Swedish examples are provided in the following chapters. *Revisionary* or reformist movements accept some of the purposes and methods of the present political order but wish to modify them. They demand reform in political democracy, property rights, class, race, religious, and gender relations. Most Swedish social movements began as revisionary collective action organizations but have evolved into conservative ones. They resist further political, economic, and social change and find it difficult to renew their goals to reflect the current needs of their members and society at large. *Revolutionary* movements completely reject the existing political order and want to supplant it with total change.

Another way to characterize collective action organizations is by their goals. Organizations which are outward- or *public interest-oriented* seek to change aspects of society and politics without considering the personal needs or problems of their members. Organizations which concentrate on changing the personal situation of their members are inward-oriented or *personal change oriented*. Many Swedish collective action organizations are a combination of the two types.

Other classifications exemplify the variety of organizations of civil society. *Class-based* organizations recruit mainly from one social class and promote and defend its interests against the interests of other social classes. The best example here are labor market organizations for workers and employers. *Non-class based* organizations recruit from all social classes and promote and defend a cause common to them. Strict public interest organizations are non-class based. They are volunteering organizations. Collective action organizations may also be comprehensive or restrictive in breadth. *Transformative* movements advocate comprehensive change in society, economics, and politics. Examples from earlier decades include the teetotaling, free church, and union movements. Today the Swedish women's and environmental movements are transformative movements. *Restrictive* movements seek change in limited areas only.

Figure 1.2 summarizes this typology of collective action organizations. Scholars use typologies to classify according to certain crucial variables or elements. Typologies tend to be ideal types. This means that certain aspects of reality are exaggerated for analysis. This book uses typologies to analyze the complex and changing nature of collective action organizations. They are useful in comparative research and make it possible to classify a particular organization in many ways. Collective action organizations may even change category over time. This book shows that this has happened to several important Swedish collective action organizations.

Swedish historians and social scientists have contributed greatly by offering definitions of social movements. They have specified a number of criteria, many of which have been incorporated into the typology of social movements and interest organizations presented in figure 1.3. The most important aspects of the ideal types involve the goals of collective action organizations. *Interest organizations* advance and defend the economic interests of their members. *Social movements* promote ideas and are mass-based organizations dedicated to promoting numerous and varying democratic values. Other criteria concern organizational structure and democracy. At first glance, these distinctions appear to be sharp and clear, but most collective action organizations are hybrids. Most collective action organizations are more-or-less social

Elements of the typology	*Typological characteristics*	
Orientation of social movement	inward (personal change)	outward (public interest)
Relationship to social class	one class (class issue)	all classes (non-class issue)
Relationship to political order	return to old order (reactionary)	status quo (conservative)
	modification (revisionist)	complete change (revolutionary)
Breadth of goals	broad (transformative)	narrow (restricted)

Figure 1.2 Typology of collective action organizations

movements or interest organizations and tend to change position over time. Examples are given in the following chapters.

The distinctions presented in figure 1.3 raise analytical questions about the true nature of many Swedish collective action organizations commonly defined as social movements. For instance, the criterion of equality among members is usually interpreted to mean equal treatment regardless of race, creed, color, gender, and social class. Yet many Swedish organizations classified as social movements have discriminated against women, immigrants, young people, and members who hold undesirable political views. The criterion of independence of social movements from their counterpart is also problematic, as many organizations Swedish social movements are highly reliant on government. Questions can even be raised about the criteria of geographic scope, member commitment, organizational size, and democratic

Social movement	*Interest organization*
Democratic (elected) governance	Professional (appointed) governance
Large membership	Size of membership varies from very small to very large
Non-economic, other-oriented goals (public interest)	Economic, self-oriented goals (self-interest)
Active member involvement in movement activities	Employed professional staff formulates organizational policy
Membership open to all	Restricted membership
Equality among members	Different member categories vary in influence
Membership as lifestyle	Membership is instrumental
Considerable geographic breadth	Varying geographic breadth
Pluralist relationship with government	Pluralist or corporatist relationship with government
Spontaneity and flexibility in member and movement behavior	Predictability in member and organizational behavior, routines

Figure 1.3 Ideal type social movements and interest organizations

governance of many collective action organizations presently classified as social movements in Sweden. Problems of classification may seem to be purely theoretical ones, but they have important political implications as discussed in chapter 8.

Political democracy and collective action organizations

Many collective action organizations seek to improve democracy by formulating and articulating the interests of specific groups and promoting them publicly. Organizations of civil society help politicians and government officials better understand the people by providing information. Autonomous and independent collective action organizations help prevent government domination over citizens and act as a political check on government and other collective action organizations. They improve public life by working together with government and other organizations. The problem is that independent collective action organizations also create serious problems for political democracy. First, organizations can stabilize political inequalities by articulating the interests of their members in adamant, ritualized, and routine ways. Discussion in Sweden on the involvement of special interests in politics has concerned this problem. Trade unions, for instance, have been criticized for opposing plans for reform of the Swedish welfare state.

Second, independent collective action organizations can deform civic consciousness. This happens when resourceful organizations dominate over unorganized individuals and small organizations with limited means for political influence. This allows them to distort and impede the expression of the general will. The impact of this problem has been felt in Sweden in the 1990s. Also, many of Sweden's largest collective action organizations have represented interests that are peripheral to their main tasks, which has hindered the development of more specialized organizations. Examples include the absence of a strong and independent Swedish consumers' and women's movements.

Third, unequal resources allow organizations to exercise disproportionate influence over public policymaking. Organizations with sound financial bases can represent their members more effectively. They can employ experts to influence policymakers and engage in expensive public relations campaigns. Today there is growing concern in Sweden about the importance of financial resources for organizations to influence public opinion and lobby government.

Finally, organizations can upset political accountability. This happens when government delegates public authority to collective action organizations and when government and organizations become intertwined and interdependent, as has been the case in such corporatist systems as the Swedish one. Citizens cannot hold politicians and civil servants solely accountable for government actions and do not choose leaders of collective action organizations in general elections.

It should not be concluded that collective action organizations are a menace to democracy. Their importance for political democracy has been made clear earlier. The underlying dilemma of political democracy is that collective action organizations are both desirable and harmful and what makes them desirable also makes them harmful. Their desirability lies in their ability to compensate for imperfections in electoral politics by formulating and ar-

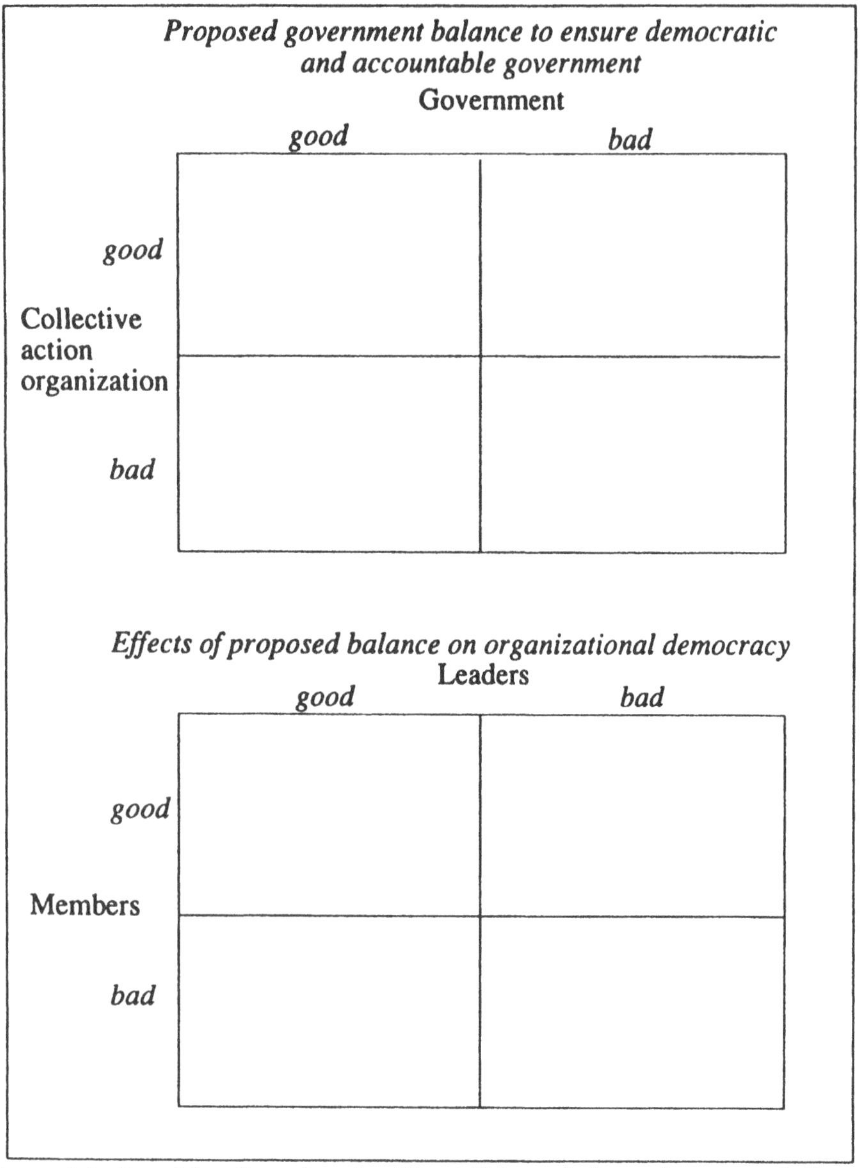

Figure 1.4 Considerations involved in striking a balance between civil society and government that maximizes democracy for all participants

ticulating the interests of citizen groups. They express intensity of preference, give voice to the less fortunate, place democratic demands on government, and provide government with valuable information. They become harmful once they lose perspective of their particular and crucial role in politics and when government becomes too reliant on them. Some scholars even blame collective action organizations for the most serious problems of modern government: atrophy of institutions of popular control; maintenance of old and creation of new structures of privilege, and conservatism in several senses of the term.

The basic problem is that a balance must be struck among three needs of modern political democracy: independent, autonomous collective action organizations of civil society; elected government to ensure political accountability to the people, and a working relationship between government and civil society to further democracy. The question is: Can democratic government be designed to control the mischiefs of special interests without endangering essential liberties and the independence and internal democracy of collective action organizations? Figure 1.4 summarizes the issues which must be considered when attempting to balance these complex demands.

Debate on the dilemma of democracy is an ongoing one, as a balance struck between civil society and government for particular circumstances, e.g., wartime, may prove inappropriate in other situations, e.g., a period of rapid reform and value change. Swedes have debated this important political issue in different periods of civil society history. A new evaluation of the balance between civil society and government is occurring today.

Pluralism and corporatism

University disciplines study collective action differently. Theories of pluralism and corporatism have dominated in political science. The theories have undergone considerable academic debate, and views vary on how they should be defined and whether they still are viable theories for contemporary analysis. This section indicates how they can be used to analyze Swedish collective action organizations.

A Swede was the first social scientist to define how corporatism functions in modern western democracies. It is unfortunate that his work has not been given international recognition. He coined the term free corporatism in 1946:

> There are different opinions on how the concept of corporatism should be defined....[T]here is general agreement that a corporation is an association of fundamentally private character which is granted special government legitimacy, possibly delegated public tasks, or controlled by government. Even opponents of corporatism must understand that this kind of relationship between the state and individual associations is present in our country and that it is

compatible with democratic government...(Heckscher, 1951, pp. 182-83, my translation).

His message was that Swedish collective action organizations are incorporated into and possibly coopted by government. Later scholars of European politics agreed with him. The theories of interest group pluralism that were formulated by U.S. scholars in the early postwar years did not take this into account. Time was ripe for a theory of government and civil society relations that considered problems of organizational democracy and the privileged status of certain collective action organizations. This was the aim of corporatist theorists.

Corporatism as a theory for analyzing western democracies began to develop in the early 1970s. It offers a realistic view of the role that collective action organizations play in society and politics in these years. Organizational leaders are seen as relatively independent actors able to discipline their members into agreement. Membership is considered as more-or-less obligatory for large groups of people, e.g., trade union membership for wage earners. This means that most members are not very committed. Scholars of corporatism have also explicitly assumed that organizational resources matter: larger, more resourceful organizations tend to dominate public policymaking. Examples of organizational resources are economic assets, large numbers of members, and staff of experts. A third important tenet of corporatist theory is that government is not only an arena where collective action organizations compete with each other for influence but an actor which promotes or legitimizes certain interests while opposing or ignoring others.

Many scholars argue that pluralism and corporatism are ideal types. This is probably the most insightful and useful outcome of the academic debate. It means that no one country is completely pluralist or corporatist and that a country can be more-or-less pluralistic or corporatist, all depending on the policy setting, organizations, and time period for study. Differences across public policy sectors reflect how integrated collective action organizations are in the government process. The two ideal typical extremes on the continuum show a pluralist, completely nonintegrated setting and a corporatist, completely integrated ones. The differences between policy sectors are captured in the concept of sponsored pluralism, which refers to situations where certain organized interests are granted access to and others excluded from policymaking settings.

Organizational resources are important here, and collective action organizations with high organizational densities are generally granted a representative monopoly by government. *Organizational density* is the term used for the level of actual members of an organization out of all its potential members. A collective action organization is a *representative monopoly* when it is considered by government and other political actors as legitimate representative of the particular interest which it organizes. Swedish examples are the Swedish Employers Association (SAF), Swedish Trade Union Confederation (LO), Central Organization of Professional Employees (TCO), and Federation of Swedish Farmers (LRF).

Both sponsored pluralism and corporatism involve integrated organizational participation in government. They differ in that countervailing powers are granted policymaking influence by government in corporatist settings. *Countervailing powers* are collective action organizations which represent counteracting or opposing interests, e.g., the LO and TCO vs. the SAF; the LRF vs. government-designated representatives of consumers, i.e., the Cooperative Union and Wholesale Society (KF), LO, and TCO.

Concern is raised about the effects of participation of collective action organizations in the political process on the ability of politicians to govern. The concept of ungovernability captures nicely this problem. Other writers are more concerned with the effects of integrated organizational participation on the interest organizations themselves, as reflected in the concepts of interest intermediation and interest inarticulation. *Interest intermediation* occurs when organizational leaders decide what the interests of their members are without consulting them directly. *Interest inarticulation* occurs when leaders cannot mediate between the interests of their members and those of their negotiating partners. The result is that the organization cannot take a stand on the issues of importance. Several Swedish collective action organizations have experienced this problem in the 1970s and 1980s.

A typology of ideal type pluralist and corporatist organizations considers their democratic framework and practices. Pluralist organizations are governed more by member values and needs. They are open organizations and not sponsored economically by government. Members are encouraged to express their opinion, and dissenting opinions and minority viewpoints are treated with tolerance and dealt with through compromise. It is easier for pluralist organizations to renew themselves. Corporatist organizations are governed more by constraints and opportunities from their institutional environment. They are characterized by elite democracy, and internal conflict within the organization is avoided by rules, procedures, and established practices. They are rigid and find organizational and policy renewal extremely difficult. Pluralist and corporatist relations have significant effects on both government and collective action organizations. Intermingling of government and civil society need not indicate a symmetric relationship. It is quite possible for government to have more power over collective action organizations or vice verse and that this power relation may differ between policy areas and change over time.

The term *decorporatization* reflects a move from a more corporatist to less corporatist system. The changes concern the relationship between collective action organizations and government as well as their relationship with their members, other organizations, the general public, and mass media. A transition to less integrated organizational participation in government and reevaluation of an organization's status as a representative monopoly are examples of decorporatization. Decorporatization implies the resurgence of civil society as an independent social, political, and economic force. Sweden is now becoming decorporatized.

The move from a more pluralist to less pluralist system is captured in the term sponsored pluralism, which signifies that access to government is less free for some organizations and more regulated for others. An important question is whether Sweden is moving towards sponsored pluralism. Some people wonder whether the present non-socialist government is orchestrating this move by cutting off contact with the LO and giving more access to the SAF. Figure 1.5 summarizes this discussion on the differences between pluralism and corporatism.

Corporatist theory is now outdated for the study of civil society and government relations. Its value lies in its perspective on the balance between government and civil society that characterized the period of welfare state growth. A few writers even asked whether corporatist relations were 'fair

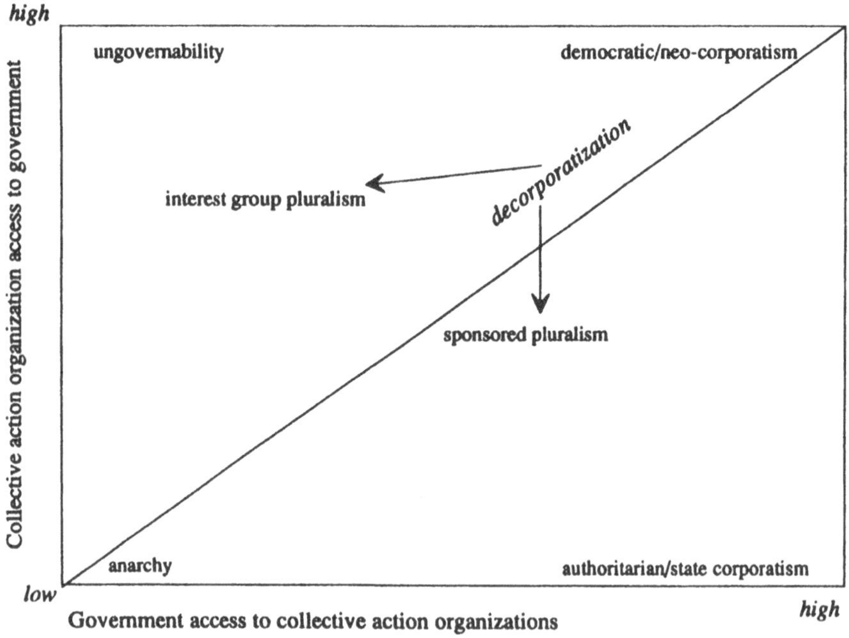

Figure 1.5 Relationship between civil society and government

weather products', i.e., where all sets of interests benefit from the exchange as government expansion ensured that negotiated settlements are not zero-sum. Other scholars wondered whether corporatist patterns fostered the emergence of new social movements, as established, corporatist organizations have difficulty in renewing themselves. Corporatist theory is obsolete in other ways as well. The role of the mass media as an arena for group conflict and a political actor in its own right has been neglected by corporatist writers. Neither has the impact of internationalization and decentralization on the relations between government and civil society been dealt with adequately.

Organizational waves and life cycles

Historians and sociologists commonly analyze the formation of collective action organizations in waves of development, and they have formulated life cycle models to explain organizational growth over time. There are good pedagogic and heuristic reasons for using these approaches. Alva Myrdal was the first Swede to discuss the emergence of Swedish collective action organizations in waves. Today we point to five basic waves of organizational development that reflect basic social, economic, and political transformation of the country. Most western countries have also experienced five waves of organizational development, but they tend to be dated somewhat earlier, which shows that it took Sweden longer to become industrialized and urbanized.

The first wave concerns the *struggle for freedom and against privilege.* It is dated to the mid-1800s. The struggle in Sweden involved peace, alcohol abuse, religious and political freedom, and gender rights for women. Liberal-oriented trade unions, the women's association Fredrika-Bremer-Association (FBF), the Swedish Peace and Arbitration Association (SFSF), temperance and teetotaling associations, and such religious groups as the Covenant, Baptist, and Methodist churches emerged in this wave. These collective action organizations protested against the political order– the establishment– of the time and wanted to redefine the meaning of political democracy. They were transformative, public interest-oriented, revisionary, and class and non-class based.

The second wave involved *struggles for economic rights and security.* The important time period is the turn of the century. The main collective action organizations to emerge, develop, and thrive were labor unions, employers' and business organizations as well as farmers' and consumer cooperatives, i.e., class-based and revisionary organizations representing economic interests. Examples include the LO, KF, SAF, and a predecessor to the LRF.

The first two decades of the 1900s were fertile ground for the formation of collective action organizations in many countries. Swedish organizations created in this period are the adult education societies, leisure time, the Swedish Conservation Society (SNF), Swedish Tourist Association (STF), and the sports' movement. This third wave can be characterized as a time of *conscious self-betterment (personal change) and citizen empowerment.*

The fourth wave began in the late 1950s, when student, counterculture, antiwar, and environmental organizations emerged. Also, newer kinds of religious, women's, teetotaling, sports and leisure time associations drew support from established collective action organizations. Early examples include the Ad Hoc Group Against Swedish Atomic Bomb (AMSA), radical women's association Group 8, and anti-Vietnam war groups. Organizations that formed later in the wave are the People's Campaign Against Nuclear Power, Greenpeace, and the fundamentalist Christian church Word of Life. These organizations are placing political demands differing considerably from those now raised by older, established ones. As such, they build upon and challenge the collective action organizations formed in earlier waves. The fourth wave may be characterized as the *struggle for personal integrity, recognition, and respect and against destruction.*

The fifth wave concerns the *need for community building,* active responsible citizens, and social solidarity. It represents a break with past waves, as it does not aim at convincing government to broaden its responsibility. Rather it aims at encouraging citizen participation in the governance of the nation. Active participation is essential here. Examples of fifth wave organizations are new and renewed volunteering and non-profit organizations. A volunteering sector was created in Sweden in the 1990s.

Figure 1.6 offers a systematic overview of the five waves. It shows that similar kinds of collective action organizations have developed in historic clusters. They tend to form when 'grievances are accompanied by new resources, expanded political opportunities, re-alignments among mass publics, and conflicts among elites' (Tarrow, 1988b, p. 283). This is the basic explanation for waves.

Collective action organizations as those which emerged in the first and fourth wave develop in times of social, political, and economic unrest or in a special kind of zeitgeist, social mood, or cultural climate that:

> creates a specific sensitivity for problems,.. narrows or broadens the horizon of what seems socially and politically feasible,...directs patterns of political behavior and life-styles,...channels psycho-social energies outward into the public or inward into the private sphere. Thus it provides or deprives social movements of essential public response (Brand, 1990, p. 28).

Wave	*Time Period*	*Aim*	*Typology*	*Example*
One	mid-1800s	freedom vs. privilege	revisionary transformative public-interest class/non-class	temperance and teetotaling societies, non-conformist churches, union, women's, and peace movements
Two	1860s to 1920s	economic rights	revisionary transformative restricted public-interest class/non-class	labor market organizations, consumer, housing, and farmer cooperatives
Three	1900-1920	citizen empowerment	personal change	adult education societies
Four	late 1950s-1990s-	integrity vs. destruction	transformative public interest revisionary non-class based	new and renewed social movements (environment, women's, peace, religious), NIMBIES
Five	1990s-	community-building, governance	public interest, non-class based	new and renewed volunteering organizations, non-profit organizations

Figure 1.6 Organizational waves of development

The systematization of collective action organization into waves of development raises a number of research questions. What happens to collective action organizations formed within a particular wave once the wave ebbs? Can organizations characteristic of one wave sustain themselves in the euphoria (i.e., upward inclines) of other waves? Do the organizations of the first, second, and third waves have a role to play in the new settings of the fourth and fifth waves of organizational development? Chapters 6 and 7 discuss these questions.

Social scientists even wonder whether unions can survive without Taylorism, a system of industrial relations from earlier decades of this century. They point to the close historical association between class consciousness and industrialism. A class conscious labor movement integrates three components: 'defense of the specific interests of workers, attack against an enemy identified through an economic situation, and active participation in the values of industrial society'(Touraine, 1986, p. 157). The components are now disintegrating, which is causing a decline of class consciousness.

Figure 1.6 should not be construed to mean that a new wave occurs when the goals of organizations comprising the preceding one have been satisfactorily attained. Rather, it shows how social, economic, and political development creates new social groups and interests which seek to solve their problems through collective action.

Life cycle models help us analyze what happens to collective action organizations once they grow older and are challenged by newer, more vibrant organizations. A four-phase life cycle model is developed to analyze Swedish collective action organizations. The model is an ideal typical construct and focuses research attention on problems that have been found to characterize collective action organizations in various stages of growth. It differs from other life cycle models in that it incorporates a greater number of aspects of importance for a complete analysis of organizational growth. These aspects include the four essential and interconnected relationships that must be analyzed in any comprehensive study of collective action organizations: (1) the organization and government; (2) the organization and its members; (3) the organization and other associations of civil society, and (4) the organization, public opinion, and mass media. Writings on resource mobilization, public choice, new social movements, pluralism, and corporatism are the theoretical basis of the model.

The first phase is *organizational genesis*. What characterizes this phase is leader and member commitment to organizational goals, high levels of participation, voluntariness, and general enthusiasm. Collective incentives, in particular purposive ones, are important. Collective incentives involve satisfaction that individual members receive from collective action. Purposive incentives are intangible costs and benefits that an individual member receives from supporting a political cause, principle, etc. Members feel personally satisfied by working for what they see as a good cause.

Phase two is *institutionalization*. Characteristics include the passage from a lively and dedicated to more professional type of participation. Individuals begin to wonder why they joined the organization and remain as members. Institutionalization means that 'the organization stabilizes, develops stable survival interests and just as stable organizational loyalties' (Panebianco, 1988, p. 18). Purposive and solidary incentives play an important role in early and selective incentives in late institutionalization. Solidarity incentives are intangible costs and benefits of a social nature that convince an individual to join an organization and remain as member. A good example is need for companionship. Selective incentives are material and immaterial reimbursements to individual members.

While charismatic leaders or entrepreneurs are important in the first phase, organizational leadership becomes more important in the institutional phase. Leaders become statesmen. Organizational problems must be solved; the organization needs to deal with outside actors and be run effectively. Institutionalization implies that collective action organizations are complex and complicated. It also signifies that the organization is gaining public and government approval and becoming legitimized. These changes affect organizational ideology in that organizations are more careful in their statements and actions. The objectives of the organization begin to become vague, implicit, and contradictory.

Maturity is phase three. Participation is professionalized, and selective incentives are important. Fringe benefits are commonly used as selective incentives in this phase. Their use signifies that members are customers of the organization. Leaders are bound by organizational precedents and traditions and function more as administrative heads for a salaried staff than leaders of a cause. The organization lacks flexibility and spontaneity, and communication problems become very evident. Member commitment to the organization is low.

Four serious problems face collective action organizations once they become mature. They must improve: (1) internal communication between staff and members and external communication with outside supporters; (2) their leadership skills; (3) the organizational program or ideology to rally supporters to the urgency of the cause, and (4) the public image of the organization.

Sense of urgency is important for collective action organizations. When it is high, supporters believe that struggles for the organizational cause are extremely important. When it is low, they do not feel that they need to exert effort for the organization. Important instruments for heightening the sense of urgency and improving communication are organizational newspapers and publications and other methods of opinion formation, and recruitment of leaders who can communicate well with organizational insiders and outsiders. Opinion formation techniques make supporters and the general public aware of the importance of the organization and the urgency of its cause.

When mature collective action organizations devote their energy to solving these four serious problems, they engage in offensive organizational maintenance or renewal. Goals are transformed and revitalized to better reflect member demands and outside forces. Discontent is taken seriously and translated into effective and concerted action more attuned to the Zeitgeist of the era. Organizational maintenance is a high priority for all healthy and mature collective action organizations. Offensive organizational maintenance even means that organizations must be very sensitive to the opinions of new generation of members. A sign of organizational stagnation, the fourth phase, is the inability to mobilize new members. Target groups for Swedish collective action organizations wanting to renew themselves are young people, women, and immigrants.

Collective action organizations that are not able to solve the four problems may instead engage in defensive organizational maintenance or superficial attempts at problemsolving. Organizations which choose this alternative usually refuse to listen to member criticism. They are oligarchical, which means that democratic power is abused by a minority of individuals within the organization. A former president of the LO offered a good characterization of oligarchical tendencies in the union movement when he coined the term 'cement butts' to characterize the proclivity of many staff officials to safeguard their employment positions more than the goals of the organization.

Organizations which do not succeed with offensive organizational maintenance decline, decay, or demobilize. The fourth phase of the organizational life cycle is *stagnation,* and it develops when an organization can no longer hold its members, its cause is scorned, and it has severe economic and structural problems. People criticize its actions, and it loses respect, legitimacy, and trust. There is a general decline in its resources and standing in society. In short, the organization is discredited internally and externally. Figure 1.7. summarizes the phases of the life cycle of collective action organizations.

Stagnation is a complex and long process and does not necessarily imply that the organization will cease to exist or disappear. It may even be difficult in the short run to know whether a collective action organization is undergoing stagnation or constructive transformation. This is the case because it cannot easily be determined whether it has or has not succeeded with organizational maintenance. Statistics on membership rates are not really a good indicator here. Long-term trends are important, and it is necessary to

		Characteristics of organization			
Phase	*Outside world's reaction*	*Leaders*	*Members*	*Incentives*	*Structure*
genesis	skeptical	charismatic	high commitment/participation	collective	loose, ad hoc
institutionalization	positive	statesmanly	partial commitment, professional participation	collective selective	tighter, developing
maturity	trustful partnership	administrator	low commitment, members become customers	selective	inflexible, routine bureaucratic with communication problems
stagnation	distrustful	oligarchical	low to no commitment and participation, members are alienated	multitudes of selective	incapability

Figure 1.7 Phases in life cycle of collective action organizations

focus attention on the ability of the organization to renew itself by recruiting new generations of members rather than by simply retaining its present members or recruiting from traditional groups.

The reason for this is the declining importance of members and large memberships for organizations in the last two phases of the life cycle. Members become less important for collective action organizations in three main ways. Funding of the organization is no longer solely based on membership fees and can be attained from elsewhere, for instance from government. Members are no longer needed as voluntary workers; employed staffs have replaced them. Mass communication makes it easier for large organizations to reach people without the help of mass memberships to pass out handouts and gather in demonstrations.

A number of internal factors are also important. Stagnation can occur rapidly or slowly. The size and age of an organization as well as its type and use of incentives will influence its ability to survive. Organizations with broader goals that use solidary and material incentives are more likely to endure. External factors play an important role here as well. Competition from other organizations over members may force the organization to engage in offensive organizational maintenance. Yet as shown in the Swedish case, support for the organization from outside may delay needed organizational renewal. Outside support may be in the form of funding, favorable treatment by government, and public approval. Many of Sweden's first collective action organizations have been able to delay problemsolving because they have lived on past achievements and glories as well as outside financial support.

Neither can success and failure at goal achievement be used as a measure of decay. Many collective action organizations exist only because they have failed to fulfill their goals: peace movements have emerged because there is no peace; women's organizations because there is gender discrimination. Success may even weaken the organization. Success often means that organizational members lose their common enemy, which leads to internal conflict over organizational goals. Extremely low unemployment and a good national economic situation, for instance, tend to work against Swedish trade unions. There are also a number of Swedish examples of organizations which seemed to have disappeared only to reappear in later periods more conducive to their objectives. These organizations have been sleeping or in abeyance. Many of the Swedish collective action organizations which emerged in the fourth wave have their roots in the first wave. They include women's, peace, and environmental associations.

Collective action organizations which are in a stage of abeyance may never really have reached the stage of institutionalization or maturity. They may not have been able to mobilize sufficient support for their causes. Initial success in support mobilization may have had more to do with ideological climate or zeitgeist. Individuals may have supported the organization for this reason only. Once the zeitgeist changes, leaders may be unable to keep up this sense of urgency and hold on to members. Supporters may join a new cause or opt for private rather than collective action.

Member commitment to the organizational cause can even change over time. This is called shifting involvement. It concerns choices about whether to become involved in collective action or use private means to solve problems. Individuals may decide to participate less in collective organizations because they are satisfied, dissatisfied, or disappointed with the results of their efforts. These feelings may range from an extreme of a pure subjective state of fatigue, stress, weariness, and anxiety to another one of an objective state regarding the inability of the group to attain its goals after considerable efforts and well-organized collective action. Between these two poles is a sense of overcommitment to and boredom with a cause and expectation-gone-wrong. Perhaps the most common experience is when an individual realizes that s/he has expected too much from collective action.

This discussion shows that the life cycle model is neither deterministic nor a natural history of collective action organizations. Organizations that reach the genesis phase do not necessarily proceed through the phases of institutionalization, maturity, and stagnation. Institutionalization may be frustrated by internal and external organizational factors, and stagnation may be arrested by offensive organizational maintenance. This does not mean that life cycle models are not important. They focus scholarly attention on the

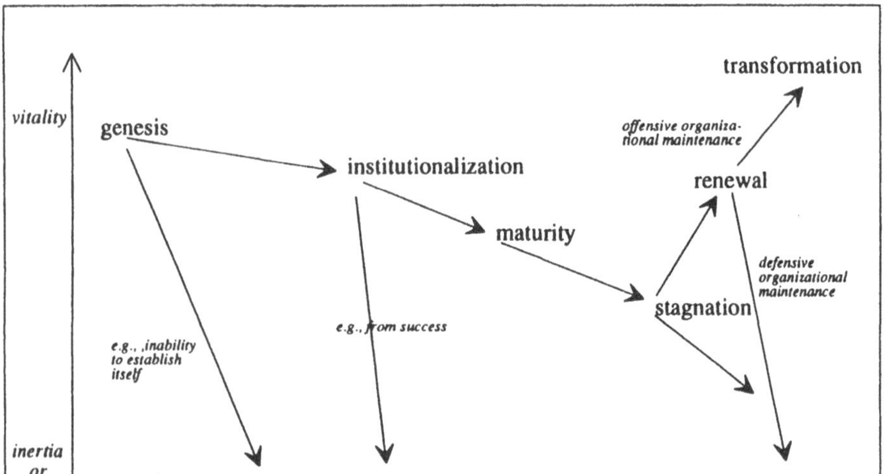

Figure 1.8 Life cycle model of collective action organizations

kinds of problems that confront collective action organizations, point to scenarios of organizational development, and help us understand organizational problems in more systematic ways. Figure 1.8 offers an overview of the discussion on life cycles.

Political opportunity structure

Social scientists stress the importance of political environment for the success of collective action organizations. This means that legitimacy of organizations is based on how well government, other political actors, and the general public receive them. The concept of political opportunity structure helps specify the environmental factors of importance and aids in comparing organizations across political borders and over time. Sweden is often compared with other countries in these studies.

With certain refinements, the concept can be used to analyze the public status of all kinds of collective action organizations. Its key aspects emphasize the impact of institutional, governmental, and political changes on the relative position of organizations and their ability to influence politics. Social scientists using the concept have been able to draw important conclusions regarding differences among countries. It is even useful for analyzing change in the public status of collective action organizations over time within a country. This is the case because all collective action organizations are influenced by: (1) structure of government; (2) legal norms and framework; (3) openness of the political system, and (4) competition among political actors. They are the key aspects of the concept of political opportunity structure. Framework of law has not really been discussed in earlier writings on the concept and is developed more fully here.

Structure of government refers to the form of government (e.g., parliamentary or presidential), organization of the central administrative process, and centralized/decentralized and unitary/federative character of the state. Nations may change their structure of government, as has occurred on several occasions in Sweden.

The *legal framework* of a political system impacts collective action organizations in particular ways. Laws are based on normative considerations, which legitimize certain values and ways of articulating and organizing interests. They may restrict the actions of collective action organizations (as done by the U.S. laws on campaign finance and labor unions) and sanction the existence of certain organizations through government financing and legitimacy (as in Sweden). Laws are important for organizational existence and strategies. They promote or prevent certain collective action organizations from developing into institutionalized, professional bodies as well as define what is conventional/unconventional and legal/illegal behavior. It is of special interest to analyze changes in laws to see how they effect collective action organizations.

Openness/closeness of political systems is the third aspect of political opportunity structures. It concerns structural access to political policymaking. Openness or accessibility may involve all stages of the political process: from agenda-creation and issue advocacy to policy formulation and implementation. The political institutions or structures generally included in the analysis of political opportunities are the executive, legislative, and judicial branches of government. Sometimes political parties are included. Accessible institutions listen to the demands of collective action organizations, integrate them in their activities, and implement them in their programs.

The access of collective action organization to government varies considerably. Certain political systems allow all kinds of collective action organizations to participate in the agenda-creating stage of the political process but prohibit them from entering more decisive stages of policy formation and implementation. Other systems may select certain organizations and allow them access to the entire process of political policymaking. Also, it is possible that the openness of a political system may change over time, a development which impacts individual collective action organizations in different ways.

It is possible to construct two ideal type political opportunity structures. *Pluralist political opportunity structures* are open to all kinds of organizations, but access is based on issue advocacy and varies over issue areas. *Corporatist political opportunity structures* give access only to government-sanctioned organizations in all stages of the political process, as has been the case in Sweden until the 1990s. A study on the women's movement in the U.S., U.K., and Sweden illustrates the importance of political opportunity structure. The U.S. political opportunity structure is classified as more pluralist than the European ones. In no country in Western Europe has a national mass-based organization similar to the American National Organization for Women emerged. This reflects the position of organized interests in the U.S., as actors outside the political process that use 'virtually every routine and non-routine method to obtain political influence' (Costain and Costain 1987 quoted in Katzenstein, 1987, p. 12). The British political opportunity structure is more corporatist. It only includes labor market organizations and limits possibilities for other groups to influence politics. It 'is organized to make exclusion from access remarkably easy' for promotional groups seeking change (Gelb, 1987, p. 268).

The Swedish structure is described as anticipatory, cooptative, and selectively open. It has been corporatist because only certain collective action organizations are allowed regular and routine access to government. At the same time it has been open to demands made by many other collective action organizations, which are incorporated into the goal structures of the established, legitimized organizations. It is more open to new ideas and values but less receptive to new political actors. This means that the boundaries between Swedish civil society and government are highly blurred and that the corporatist relationship between the labor market organizations and government has greatly influenced the character and development of other collective action organizations. Non-class and non-labor market collective action organizations have found it difficult to gain public attention. This is a consequence of a highly active, interventionist corporatist state and historically stable hegemony of social democracy.

The fourth part of political opportunity structure is *competition*. It involves divisions within the political elite and elite tolerance/intolerance of protest. Disagreement may signify that certain political leaders choose to mobilize support for their own point of view, which opens up political opportunities for collective action organizations. They may even leak information to outsiders, which can be used by organizations to gain access to the political system. Elite disagreement may also lead to collective action, i.e., formation of collective action organizations. Competition among political actors is a central aspect of the political opportunity structure concept because it is assumed to lead to access by weakening hegemonic control and forcing political actors to seek new bases of support.

Stability/instability of political alignments concerns the general issue of political hegemony and more narrow questions of political party strength and vote maximization. Political parties experiencing low popular support usually give access to new social groups. Political parties legitimate issues and attribute public status to collective action organizations. The electoral instability in Sweden, which began in the 1970s, is creating new opportunities for political access for women and other groups. Stable/instable political alignments also affect decisions on how influence can most effectively be exercised. Frequent changes in government or divided or minority governments, which must secure legislative support for their programs, force organizations to adapt methods of influence suitable for turbulent political conditions. Lobbying is one alternative. Flexibility is extremely important here because traditional and routine contacts with one political party may prove detrimental. A collective action organization that identifies itself or is identified with a particular political party may find that it loses much of its political influence once the party's electoral support declines. An important research question is whether instable government arrangements cause collective action organizations to become less committed politically, inclined to contact and cooperate (i.e., build up networks) with a greater number of political parties, and eager to diversify their methods of influence. A third aspect of competition is the presence or absence of allies and support groups, an important indicator of the status of a particular collective action organization. Allies may be other collective action organizations, political parties, and even government.

Comparative research on the women's movement in the U.S. and Western Europe offers thought-provoking results on the importance of allies and support groups. It shows that women's organizations tend to seek allies on the left of the political spectrum, e.g., the Democratic Party in the U.S., Labour Party in the U.K., and labor movement in Sweden. Government centralization and strong trade unions have influenced other organizations as well. A key issue for research on contemporary Sweden concerns the consequences of government decentralization and decline of trade unions and social democracy for the structure of political opportunities and character and development of Swedish collective action organizations. Will the new Sweden continue to anticipate the demands of new groups and incorporate them into routine politics? Or will collective action organizations be forced to adapt themselves to a completely different structure of political opportunities that required clearer and sharper divisions between civil society and government? What is going to happen now that government is decentralizing, social democratic hegemony waning, and labor market organizations stagnating? A reasonable guess is that lobbying will be used even more by collective action organizations.

Lobbying became more prevalent in Sweden in the 1970s. It is a type of opinion formation activity that is used in countries with more pluralist opportunity structures. When a representative for a collective action organization contacts government decisionmakers s/he is engaged in direct lobbying. Public relation lobbyists work with indirect lobbying or image building and modification. Its targets are elite groups and the general public. Modern communication has enhanced the importance and effectiveness of indirect lobbying as a form of opinion formation activity. Computers manufacture personalized letters to recruit support for collective action organizations, and media campaigns create goodwill for the organization and communicate its message to the general public. Media campaigns may be offensive and promote a specific policy objective or defensive and prevent changes in the existing order. It is common for well-established organizations to practice defensive media propaganda. The emergence of indirect lobbying is just one example of the importance of devising new methods for influence.

A common lobbying practice is interest group ratings, i.e., published evaluations of how political parties or individual politicians vote on legislative proposals of importance to the collective action organization. Interest group ratings were first used in Sweden in the late 1800s, but change in the political opportunity structure made them obsolete. They became popular again in the late 1970s and are now used by many organizations to mobilize support for their causes. Think tanks are used to produce information for lobbying endeavors.

The terms opinion formation, persuasion, lobbying, and pressure group politics characterize the methods used by collective action organizations that lack intimate, reliable, and stable ties with government. Such methods are less important in countries where collective action organizations occupy authority roles, consultation is regularized, and government access is ensured, as has been the case in corporatist settings where bargaining or an exchange of power has occurred between selected collective action organizations and government.

Changes in the corporatist political opportunity structure– for example through decorporatization– imply that collective action organizations must reconsider their position in politics. Decorporatization creates strategic and tactical problems for collective action organizations. They must revamp their methods of action and adapt to a more pluralist setting. Collective action organizations that evolved in corporatist settings have developed corporatist goals, procedures, structures, and traditions. The process of decorporatization forces them to rethink their traditions. This may lead corporatist organizations to experience an identity crisis. This seems to be happening to established Swedish collective action organizations.

New social science view of Swedish civil society

Perspectives on collective action organizations have developed greatly over the last century. This is explained by the changing nature of collective action organizations and increasing inclination for social scientists to use comparative approaches. Different time periods are compared with each other to understand similarities and dissimilarities in collective action, and corresponding organizations from varying countries are analyzed side-by-side to explain national differences. There is also greater exchange of ideas and approaches among social scientists. Persuasive arguments by economists have influenced how social scientists from other disciplines view collective action organizations, and sociologists and political scientists have begun to develop a common pool of research perspectives and methods. More scholarly work is needed to bring social science theories up-to-date with current reality.

There are a number of important research questions that need to be addressed in any serious study of collective action organizations. They are summarized briefly here and have been used as analytical guidelines in this book.

(1) It is possible to conceive of different relationships between collective action organizations and government. These relationships may vary over time, among organizations, and policy areas. This book analyzes the differences and changes in the Swedish relationship by focusing on the impact of social, political, and economic developments on perceptions of the proper role of government and civil society. Relevant developments here are demographic changes, technological advances, emergence of new social groups, and impact of international events and situations.

(2) Swedes become interested in collective action organizations for different reasons, and their involvement may vary in intensity and over time. Their involvement may be connected with cycles of collective action and organizational incentives. Membership should not be taken as a simple expression of commitment to collective action organizations. It is a complex phenomenon that deserves sophisticated analysis.

(3) Collective action organizations in Sweden are not only dependent on members for support. Outsiders are necessary sources of organizational resources. Outsiders may be the mass media, other collective action organizations, public opinion, and government. To be successful, collective action organizations must mobilize support from members as well as from their political environment.

(4) Collective action organizations are engaged in the debate over the meaning of political democracy. They may defend or challenge the prevailing political order, and their challenges may lead to a widening or a narrowing of the boundaries of the political. They are part of the process of rethinking and reevaluating Swedish political democracy.

(5) The organizational life cycle offers a good point of departure for analyzing changes in organizational character and the general problems facing Swedish collective action organizations in different phases of their development. It is important for organizations which have reached maturity to be able to renew themselves. This process is called organizational maintenance, and an important distinction is made between constructive revival or offensive organizational maintenance and protective, status quo, or defensive organizational maintenance. Collective action organizations which are highly dependent on government find it difficult to adapt themselves to new circumstances. They are tied to the old political order and threatened by everything that is new. The inability of older, established organizations to rethink democracy and adapt to the Zeitgeist is one explanation for the emergence of new waves of organization.

(6) An important research topic is the position of Swedish collective action organizations on the political order. There are four ideal type positions here: reactionary; conservative; revisionist, and revolutionary. Old social movements defend and support the present political order; they are conservative

in orientation. New social movements struggle to reenact democracy and rethink politics. They are reformist or revolutionary in orientation because they want to change the boundaries of the political to broaden democracy.

(7) The importance of political environment for collective action organizations is emphasized in the discussion on political opportunity structure. Key aspects are structure of government, competition among political actors, openness of the political process, and legal norms and framework. The political opportunity structure of a given country influences choices made by collective action organizations. Public opinion formation and particularly lobbying are used in pluralist settings, while corporatist organizations rely more on the exchange of power with government. This book examines how Sweden's political opportunity structure has changed and impacted the structure and activities of collective action organizations.

(8) Study of collective action organizations entails analysis of four essential relationships: (1) the organization and government; (2) the organization and its members, (3) the organization and other associations of civil society, and (4) the organization, public opinion, and the mass media. Full understanding of the subject matter requires that the four relationships be seen as interconnected. Analysis of their interconnectness is particularly important in historic times as the current one, when many of our theories are proving to be inadequate and outdated. This book shows how traditional theories and new thinking within social science can be combined and constructively applied to analyze the transformation of Swedish collective action organizations and changes.

2 Formation of modern Swedish civil society

Sweden underwent extreme transformation from the mid-1800s. Swedish demography changed greatly when a law was passed that permitted parcelling out of land. This land reform led to destruction of old Swedish society, with the state church as the village focal point, and displacement of vast groups of people, through what would later be called unemployment. These changes led to various kinds of social misery. The society of old Sweden– built on the authority of the state church and other patriarchical relationships– crumbled and, for a while, it did not seem that any other kind of social organization would replace it. The social stability which had been created by these relationships and institutions was disappearing rapidly. The country was developing into a mass society, but this was checked by the emergence of collective action organizations, in particular the free churches and temperance and teetotaling societies. This chapter analyzes collective action organizations in the first period of modern civil society history from the 1850s to 1890s.

Free church movement

Research has shown that religious movements are usually the first group to emerge in tumultuous times, and given the Swedish development, it should come as no surprise that the first inklings of social movements on the Swedish scene were new churches. In fact, implementation of the land parcelling-out reform and emergence of free churches are positively correlated, and free churches flourished in those areas where land reform was most successful. The incentive to establish a new religion came from abroad. Some of the most successful organizers were foreign, coming mainly from England and the

U.S. The organizers traveled throughout the country and propagated for the new cause. New Swedish adherents felt that they were part of an international movement:

> A Baptist in an one-horse town felt that he no longer was a simple tailor or crofter. He was part of a world-wide army of millions that could even report continuous victories in the struggle. Self-confidence in non-conformist religious thinking was thereby strengthened. New songs, new forms for meetings, and new attitudes were brought into the Baptist chapel and mission house by Swedish Americans returning home. The golden crowns on their teeth and the 'w' in the name 'Swanson' added an extra touch to the excitement surrounding their foreignness (Sundström, 1952, p. 65, my translation).

The political order of the time viewed the followers of the first non-conformist or free church religious associations as deviants and threats to society. Even their clothing and general demeanor was different. Some followers were exiled; others left Sweden voluntarily, though a few returned instilled with the religious freedom of the new world, and many of those who remained were persecuted by the state church and prosecuted, fined, and imprisoned by government. Their actions and beliefs were illegal; they were true political deviants. Reaction by the established political order to these new ways of thinking and behaving was very harsh. Clashes between the establishment and its challengers could, on occasion, become violent. Rules and regulations which had laid dormant on the law books were invoked to squelch the new sentiment which threatened the protectors of established nineteenth century political, social, and religious order. One of the first developments toward religious freedom, a weak law from 1860 which allowed individuals after much effort to leave the state church, came too late to check the new, defiant religious sentiment.

A general Swedish pattern of behavior which involves disdaining the ideas and behavior of new social and political actors developed at this time. As will become clear later, this pattern of behavior has repeated itself time-and-again. The cast changes in that those actors, who once challenged the established political order, become over time its most adamant defenders. They condemn the goals, methods, and activities of newer social and political actors.

The situation grew into a struggle for freedom and against privilege and, as is often the case with general struggles of this kind, it led to unusual coalitions of supporters. The foreign press condemned Sweden for its religious intolerance, and the Swedish liberal press, which definitely could not be considered as a friend of religion, criticized the state church for its repressive measures that defied individual freedom of speech and association.

The newly-formed religious societies practiced pietism, a doctrine of religious and radical individualism which formulated new criteria for Christian women and men. They challenged established authority because they believed that each and every individual was able to interpret the will of God. Emphasis was put on the teachings of God rather than the authority of priests and the state church. Missionary work became an important area of activity. Followers came from all walks of life, but the most determined members came from the new middle class of the time.

This new thinking was even manifested in the architecture of the first religious meeting houses and chapels constructed in the 1850s and 1860s. They were simple rooms plainly decorated and frequently without any exterior religious symbolism. Unlike the state church, there was no segregation in seating arrangement, which meant that followers from all social classes mingled together. The new religious associations had different reasons for their choice of architectural design. One low church, the Evangelical Fatherland Foundation (EFS), was reformist, considered its services as complementing rather than defying those of the state church, and showed its position by constructing buildings that could not possibly be mistaken for churches. The Baptists were, in contrast, more revolutionary and wanted to replace the state church. They were the most deviant and created considerable worry for the political order, and many of its leaders and supporters were exiled.

This is one illustration of the relationship between the first social movements in Sweden and the established political order. Also, it shows that a social movement is composed of different organizations which do not necessarily agree on every policy detail. In fact, certain free churches began as factions within low or other free churches and broke away when disagreement became intense. Some groups within the movement were revisionary while others were more revolutionary. It is noteworthy that the low and even the first free churches– the Covenant, Baptist, and Methodist Churches– did not completely break with the basic theology of the state church, nor did their members leave it. The break came when the Pentecostal movement and fundamentalist churches emerged in later periods. The membership rates of Swedish free churches are found in appendix 1.

Later the free churches gained popular support and became legitimized public actors. Free churchmen became members of parliament and representatives in regional and local governments. An association for parliamentarians of free church faith formed in the late 1880s and exists today. Financial support from government to the free churches was provided as well. This tradition continues to the present day. The negative position taken by the political order towards the free churches in earlier years changing dramatically and rapidly. It was, for example, possible for a Baptist to become a MP only sixteen years after a Swedish court ruled that another Baptist be exiled for his religious beliefs. Change in attitude on the part of the political order once a social movement has established itself is a second general characteristic of Swedish civil society history.

Temperance and teetotaling movements

The temperance movement is the second important social movement to emerge before Sweden became a political democracy. This movement started because alcoholism was a serious problem in the nineteenth century. The first temperance association was founded in 1819, but the temperance movement took root in the 1830s. These 'moderation' societies only condemned consumption of distilled alcohol. They were religious in orientation, and their first leaders came from high society. Many methodists were involved in the movement. The two foremost foreign organizers of the free church movement also actively promoted the establishment of the first Swedish temperance societies. For them, religion and temperance went hand-in-hand. This relationship is probably the first instance of a social movement network in Sweden. By 1845 the national organization had around 100,000 individual members and over 400 member associations. Nordic meetings were held. These were the best years of the temperance movement. It was in the genesis stage of its organizational life cycle.

The societies concentrated on educating the public about the sinfulness of alcohol. Movement entrepreneurs believed that public opinion formation would lead to changes in drinking patterns and government policy on alcohol. The movement was successful in this regard. It persuaded politicians to enact a law on distillation in 1855. 'Oral propaganda' (traveling speakers, all of whom were priests) was an essential part of this activity. The movement understood the power of numbers and engaged in what was called a 'petition' or 'opinion' movement. Over 800 petitions were submitted to the King and Estate Parliament in the early 1850s. Other Swedish social movements used these methods of action in later years. Dynamic, charismatic, and clever leadership was crucial. The temperance societies were highly instable organizationally, and the general membership lacked the necessary political resources– education and organizational talents and experience– for them to take responsibility for the societies. Many members came from the lower echelons of society; recruitment to the movement was particularly successful among the rural underclass.

The movement began to decline before it became institutionalized. Its success with the enactment of the 1855 law is one explanation. Other important reasons are general inexperience with organizational life, the decision by the main movement leader and entrepreneur to step down from office, loss of members to newer groups, problems with government authorities, and criticism by the state church of the 'Anglo-American-style tendencies' of the movement (Eriksen, 1990). Also, there was internal tension on the issue of abstinence: a fundis faction was challenging the realos views of the leadership. The temperance movement gradually died and was replaced by the teetotaling movement, which began to grow in the 1870s. Policy change did not come peacefully, however.

The teetotaling movement differed in significant ways from its predecessor. More followers and leaders came from different social classes, though many of the first leaders were highly educated men. This change signified that more groups of people were given a chance to influence society. For this reason, the teetotaling movement is considered a school of democracy. Brotherhood– not sisterhood– was a fundamental concept, and this meant that the goals of the movement were broader. Abstinence or teetotalism was demanded of members.

It was becoming a true social movement. It mobilized large numbers of people and gradually was legitimized politically. Membership rates are given in appendix 2. Many politicians were or became teetotalers. The political opportunity structure of the time made it easy for the movement to get its men nominated for and elected to representative bodies. The movement used its newspapers to publicize its positions on political candidates, with the explicit purpose of persuading its suffraged members how to vote. Today this activity is called interest group rating. A parliamentary group of teetotalers was also established and exists in different form today.

The movement organized demonstrations which attracted thousands of supporters, sent petitions calling for prohibition with tens of thousands of signatures to the Swedish King, and mobilized support by calling national meetings, which were held in election years to the Lower Chamber of Parliament. An unsuccessful teetotaling political party was founded in 1890. An important activity for each emerging teetotaling organization was the establishment of a newspaper.

The organizational idea and many influential orators came from the U.S. In fact, most of the teetotaling organizations were U.S. imports. Of them the Swedish Grand Lodge of the International Order of Good Templars (IOGT) was the most successful: it won an immediate response in Sweden and 'simply swallowed up a great deal of the 1870s religious absolutist temperance interest' (Eriksen, 1990, p. 78). The reason for the presence of so many social movement organizations was policy differences. An important one was religion. There were religious-oriented teetotaling associations, such as the IOGT which included Methodist rituals in its activities, and those which were neutral on the religious issue. Other lines of conflict involved internationalism, class discrimination, and the cause and consequences of alcoholism.

The Workers' Temperance Society (NVO), more commonly called Verdandi, began as an IOGT lodge with friendly relations with the emerging social democratic movement. It organized workers in Stockholm and later was expelled from the IOGT because it concentrated on social reforms and universal suffrage rather than the religious aspect of alcoholism. Its slogan was 'a sober proletariat, that is what the ruling class fears' (Hurd, 1990, pp. 102-3). Another organization emerged from a dispute within the IOGT, whose root was the race issue in the U.S. branch of the IOGT. Swedish followers of this faction decided that they did not want to be governed by a foreign (i.e., international) board and voted to reject members of foreign birth. They stressed the national character and Swedish roots of the teetotaling movement; emphasis on nationalism by Swedish social movements would become more prevalent in later periods.

Different policy goals even meant that the associations became specialized. The White Ribbon Society (created in 1900) concentrated on practical social issues involving women and established children's homes and schools for homemakers; it had close contact with the women's movement. Another organization started the first Swedish teetotaling life and health insurance societies. The only important issue that the movement organizations agreed upon was prohibition. It was their main policy demand from the late 1890s to 1922, when a national referendum on the issue was held.

Union movement

The trade union movement is the third important social movement for this period. The roots of the movement are from the 1850s and international. Initial mobilization was stimulated by the 1848 February revolution in France– whose fervor reverberated in violent street demonstrations involving the military and led a Swedish newspaper to capture the event in Stockholm with the words 'Revolution in Paris, Disturbances in Stockholm' (Lindorm, 1961, pp. 172-74)– and the Thraniter, utopian movement in Norway. Also, the development of a capitalist economy in Sweden changed labor relations and created class consciousness. Workers were no longer forced to accept employer patriarchy once labor became a direct economic transaction, and employer responsibility towards the employees stopped at the factory gate.

Swedish society was open and responsive at this time to significant events in other countries, and the labor movement, not unlike the free church and teetotaling movements, emerged and developed in this international wave of liberalism, industrial capitalism, democratization, and individualization.

Changes in Sweden's political opportunity structure in the 1860s generally made it easier for social movements to establish themselves. They include a law on freedom of trade which legalized union organizations, new laws for local government, and a reform that abolished the four estates and created a bicameral parliament.

An important decade for Swedish unions is the 1870s. The first unions to emerge in this decade were not part of the social democratic movement but were liberal in orientation. It was not uncommon for people from the middle class and employers to start unions, whose initial aims were rather limited in comparison with those of the social democratic labor movement soon to develop. The liberal unions started self-help schemes similar to those provided by the old craft guilds. The cutting edge of unionism was skilled craftsmen; typographers were at the forefront. Workers in the capital city of Stockholm were among the first groups to unionize. Membership was initially quite instable, but there existed a core of committed members.

The First Socialist International was founded in 1864, and its ideas spread from country to country and reached the ideologically fertile ground of Sweden in the 1880s. The Second International from the late 1880s inspired socialist movements to protest and demand, in a yearly protest every first of May, an eight-hour work day. Social democrats began to compete with the liberal unions over sympathizers and to challenge the authority of employers and the capitalist system in these years. Workers and unions were ready for this ideological current.

The first large strike occurred at a sawmill in Sundsvall in Northern Sweden in 1879. This was not a union strike; it failed badly. The employers punished the strikers by blacklisting them and evicting them from company housing. This led to worker anger, which was channelled into recruitment efforts benefiting both liberal and social democratic unions. Workers held strike meetings wherever they could, but it was not unusual for establishments to refuse workers sympathetic with social democracy access to meeting halls and spaces. Such problems led social democrats to construct their own community centers, which are literally called 'people's houses'. The inspiration here came from Belgium and France.

The social democratic trade union movement used many of the same tactics as the teetotaling and free church movements. They too put much effort into starting newspapers and other publications, and gatherings of sympathizers was important for them as well. Their efforts proved to be successful, but they still were deviants in the eyes of the political establishment and the other two social movements.

Social democrats were very successful unionizers. They had a better set-up than the recruiters for the liberal unions. Their newspapers offered good coverage of union activities, supported workers completely, and advertised for financial support for strikers. Social democratic agitators became readily involved in strike activities, and the party itself was committed to the struggle for economic rights. In contrast, it was very difficult for liberal politicians to offer workers the same kind of support.

One of the most important years in the history of Swedish unionism is 1898, when the peak organization the Swedish Trade Union Confederation (LO) was founded. This was nine years after the founding of the Swedish Social Democratic Party (SAP). The establishment of a peak union organization was debated in the party for several years. Party members argued that the labor movement should be divided into two separate organizations, a party and a union, as it was feared that union matters were becoming too dominant. There was an intrinsic partnership between these two collective action organizations from the start. The cornerstone of the party was unionism. The first president of the LO was a party man, an intellectual with no link with the working class. The relationship between the party and union has generally been friendly and stronger than in many countries, though this does not mean that the labor movement has not had its 'wars of the roses'. A term has even been coined for this relationship; it is collective affiliation.

A specialization in goals occurred after the founding of the LO. The party began to concentrate more on political issues and leave union matters to the LO. From the beginning, the LO was weakly centralized; most influence and power laid with its unions. The LO grew in strength and gradually dominated over the liberal unions. It won a number of battles, but its initial strategy was defensive. Later it would turn its attention to offensive tasks, among them the struggle for discursive power and political hegemony. Appendix 3 offers information on Swedish union membership.

Women's and peace movements

Swedes consider the free church, teetotaling, and labor movements as classic Swedish social movements. They created modern Swedish civil society by mobilizing people into associations that were not only independent of the state but initially defiant of it. They carved out a public space where people could exercise political freedom. Two other social movements emerged that deserve a place in this Swedish hall of fame. They are the women's and peace movements.

Three developments led to the establishment of a Swedish women's movement: internationalization, industrialization, and emigration. The spirit of the French revolution and its commitment to human rights inspired some political theorists, including a Swedish one, to demand the same rights for women. Industrialization signified new employment opportunities for men and women. Farmers' wives who had worked as farm labor sought employment in industry. More men emigrated, and this meant a surplus of unmarried women without employment and many citizenship rights. Increasing numbers of unwed mothers and the need for better schooling for women to prepare them for working life were two pressing social problems from the 1860s onward.

The first mobilization effort was conducted by Fredrika Bremer, an internationally-recognized and well-traveled author of novels in the mid-1800s. Swedes were aware of the significance of her books, which captured the zeitgeist of the mid-1800s. Her books dealt with the plight of women, and one book, *Hertha*, became the rallying cry and platform for the emerging women's movement. She wrote it after returning from a trip to the U.S. It was a masterpiece of opinion formation, the first open revolt of a 'middle class daughter', and has been likened to *Uncle Tom's Cabin*. Decades later the title would be used as the name of a newspaper published a women's movement organization.

The Fredrika-Bremer-Association (FBF) was founded in 1884. Though upper class women and men founded the FBF, its issues concerned the plight of all women in Swedish society. It encouraged women to use their limited rights as citizens to vote in local and regional elections and seek election to school and poor relief boards. The basic organizational concept was that women had special qualities that gave them a particular societal call. Fredrika Bremer called this 'social motherhood' and believed that 'no true democracy is possible without the complete participation of women' (Wahlström, 1944, pp. 93-94, my translation).

The association held courses and study circles on various subjects, whose aim was 'citizenship through knowledge' (Åkerman, 1944, my translation). It established associations for women workers, which today would be called networks. It organized conferences for other women's movement organizations emerging in this important period of Swedish civil society; joint statements and actions were taken. The FBF also played a role in the international women's network. Opinion formation became an increasingly important area of activity: care was given to factual presentations; a newspaper was started, and contact with politicians or lobbying was given priority. Sympathizers of the women's movement, many of whom were involved in the peace and teetotaling movements, were encouraged to motion in parliament for legislation on the rights of married women and other issues.

The women's movement did not expand at the same rate as the other ones discussed in this section. Why was this the case? One reason is the difficulty of getting the women's question on the political agenda after the 1880s. It was overshadowed by such issues as free trade, defense, and the union between Sweden and Norway. A second explanation is the successes of the early pioneers in the 1880s, a decade of legislative reform in this field. Success has not always had a positive effect on collective action organizations. Another important reason is the mistrust which met the movement. The FBF found it hard to create a positive image among women. The fact that the association was led by upper class women and men probably made it difficult for lower class women to identify with it. A women's movement sympathizer evaluated the status of the women's movement in these words:

> Compared [to the labor movement], the women's movement has been a sail boat on the surface. A small number of women have represented it, but it has never been able to create general solidarity. It has aimed at particulars, not at the whole agenda. For its victories it must give foremost thanks to general societal development, for the small group of brave struggling women would never alone have been able to win these accomplishments (von Konow, 1944, p. 10, my translation).

Class not gender solidarity was already the top political priority of Swedish civil society. The country was not really ripe for an independent women's movement. In fact, there was not much understanding for the women's question. Class solidarity was gender-biased towards men: women were generally discouraged from joining unions, and their unions were ignored by male union leaders. Internal problems also hampered the women's movement. For instance, the leaders of the FBF neglected young women by not encouraging them to assume a more active role in the movement. This has not been an uncommon experience in Swedish civil society history, but disregard of youthful strength and enthusiasm has always led to problems with member recruitment and organizational renewal. Finally, the association was too reliant on receiving attention from the press rather than creating its own publicity: the lesson for the future was that public relations should be given higher priority. Appendix 4 contains information on the membership rates of Swedish women's movement organizations.

The peace movement developed in a different way. The first existing associations were based in U.S. Quakerism. The largest movement organization, Swedish Peace and Arbitration Association (SFSF), was founded in 1883 when concern for military expenditure was a hot item on the political agenda. The SFSF is similar to other Swedish social movements in that its founders were important members of society but is different because all its founding members were MPs. Initially the SFSF was an elite organization with no ambition to become a social movement. Activities on the grassroot level were not encouraged, and this, together with waning public interest for the military expenditure issue, made it difficult for the organization to become institutionalized. Membership rates dropped, as shown in appendix 5. This situation changed in 1888 when a new president, an elementary school teacher, took over. He turned the SFSF into a social movement organization focused on public opinion formation.

There were ideological ties among the social movements that emerged in this period. Many of the programs adopted by them referred to the issue of peace. Most members were either liberals or social democrats, and this common political base strengthened the peace movement and explains the success of the SFSF in influencing Swedish politics in later years. Peace sympathizers in parliament had organized into a group in 1892 very similar to the ones within the free church and teetotaling movements. This group was the closest link the movement had with the political order. The SFSF lobbied sympathetic politicians and, as with the other social movements, encouraged them to motion on issues of importance to the movement.

Peace became again an important issue on the political agenda when the union between Sweden and Norway showed signs of strain. Many people feared that war might break out, and this led to an upswing in membership between 1895 and 1905. The SFSF stressed Nordic contact, liberal internationalism, and arbitration as a way of solving this pending conflict. Few other peace movement organizations emerged in these formative years. Peace associations for women and other collective action organizations were founded, but the peace issue did not create the same enthusiasm as the union, free church, and teetotaling movements. It was initially difficult for the movement to attract charismatic leaders, and this gradually led to movement decline. Also, its goals were more diffuse than those of the other social movements. As a consequence, the peace movement has and still is more dependent on external events to mobilize support. This explains the proclivity of scholars to discuss the movement in terms of mobilization peaks and declines. Two of the eight peaks occurred in this period (Lindkvist, 1990).

3 Sweden's democratization and social movement institutionalization

The formative period of Swedish civil society ended by the early 1900s. By the second period dating from 1900 to 1920s, each social movement had its own agenda, but they all agreed that society needed to be reshaped were they to fulfill their goals. This joint understanding led them to cooperate to change the boundary of the political. They set out to do this in various ways.

Universal suffrage and Sweden's new political opportunity structure

Different strategies to reform Sweden were initially chosen by social movement activists. A significant number of unionists believed that revolution was necessary and, together with many social democrats, they organized a political strike in 1902 during a parliamentary debate on the suffrage issue. The strike awakened the working class to the usefulness of collective action and its potential as a disciplined, mass organization. It was more a demonstration than a strike: 65 percent of the striking workers had permission from their employers to participate in it (Westerståhl, 1945, p. 224). The strike was the first mass demonstration of the labor movement in Sweden.

Other social movements were far less enthusiastic about using this tactic to broaden the boundary of the Swedish political. Most abstainers regarded the self-help and individual revival strategies of the teetotaling movement as sufficient and insisted on remaining politically uninvolved. The reaction of an old-timer sums up this position well: 'Political democracy may be good', he said, 'but what does it have do with the teetotaling movement?' (quoted in Svensson, 1979, p. 223, my translation). The free churches and the women's movement were concerned about political democracy but did not believe that striking was the proper tactic. Gradually all social movements chose the reformist route to change Sweden. They concentrated on parliamentary politics, the evolving political parties, and the issues of prohibition, labor reform, and extending suffrage to all adult citizens. Prohibition and labor reform are discussed later.

Demands for universal suffrage were not new in Sweden. An ad hoc campaign for suffrage began to develop in the late 1880s. Franchise societies were founded throughout the country, and they attempted to form public opinion by lobbying government and demonstrating. An early publicity event was the 'people's parliament', an idea imported from Belgium. Two such street parliaments were organized in the 1890s. The suffrage campaign participated actively in general election campaigns in the 1890s but began to decline shortly thereafter and died in 1900.

Swedish collective action organizations were involved in the suffrage campaign. Not every social movement supported the issue with equal force or stressed its importance for both genders. The free church and women's movements were concerned about women's suffrage, but the free church movement never became an active participant in the struggle and its interest in the issue declined over time. The labor movement was the only Swedish social movement with suffrage on its agenda, though it put more emphasis on enfranchisement of men. Officially the teetotaling movement, with the exception of Verdandi and then the IOGT from 1918, did not take a stand on universal suffrage, but many of its members were sympathetic to the cause and active in it. There were also regional variations within movements on the issue.

Attitudes on the suffrage issue coalesced into two basic political camps. The left advocated suffrage reform; it consisted of social democrats and liberals. The conservative right opposed electoral reform, as did the Upper Chamber of Parliament. Suffrage was a highly controversial political issue in these years. After the liberals were forced out of office on the issue and replaced by a conservative government in 1906, the labor movement began to debate political tactics and the pros and cons of another political strike, which was never held. By 1909, the Swedish electoral system underwent significant change. Adult male citizens were generally granted the right to vote, and elections to the Lower Chamber of Parliament were carried out in multimember constituencies on the basis of proportional representation.

The women's movement devoted almost all its attention to the suffrage question, which meant that it neglected other important issues. A special organization, the National Association for Women's Suffrage (LKPR), was established because the universal suffrage movement from the 1890s and the Social Democratic Party gave priority to the enfranchisement of men. The FBF played an instrumental role in its foundation. The LKPR was a network organization for a number of local groups. New middle class women– school teachers and office workers– dominated its ranks. Single and urban women were prevalent; it was difficult for the movement to reach religious, conservative, and working class women. Working class women had started their own suffrage association in the 1890s, but when the Social Democratic Party decided at its congress in 1905 that demands for women's suffrage were an untimely luxury, they loyally followed the platform of their party. Other groups of women remained passive in this struggle. Few– if any– Swedish women were suffragettes in the U.S. or English sense. Women's organizations lobbied government in every form imaginable and finally won the suffrage battle in 1921, years after many other countries.

All proposals on electoral reform were hotly debated, and collective action organizations carefully considered how constitutional reform would affect their ability to exercise political influence. They wondered about the effects of proportional representation and the more dominant role to be played by political parties on their methods of action and role as political actors. Two good examples of the effects of changes in Sweden's political opportunity structure come from the teetotaling and women's movements.

The IOGT saw proportional representation as worrisome for it feared that it would lead to decreased political interest in the issue of abstinence. Proportional representation also worried liberals, and it is not surprising that thinking was similar here, as many IOGT members were liberal sympathizers. The strength of the teetotaling movement was built on a number of factors, all of which began to disappear as Sweden's political opportunity structure underwent change. Majority elections in single member constituencies had favored the movement as it could influence candidate selection. Development of a party system of government signified that candidate selection would be conducted within the parties. As a consequence, they would be less accessible to movement influence. Also, movement sympathizers had, thanks to suffrage reform, left the ranks of the politically poor. Their interest in politics could be satisfied through other channels, and priority was given to the emerging political parties (Johansson, 1947).

The political opportunity structure that emerged after the first suffrage reform created problems for the LKPR as well. It wanted to retain its reputation as a nonpartisan organization but found this difficult once the social democrats and liberals took a clear position for and the conservatives against women's suffrage. Another problem was that it did not have much political influence because party loyalties were strong and gender solidarity very low among women. Once women's suffrage was realized, members and leaders of the LKPR discussed the role that their organization could play in the future. Three options were formulated: it could cease operating; continue existing as an independent association, or merge with the FBF. The LKPR folded in 1921 and was replaced by the Swedish Alliance for Women for Suffrage and Equal

Citizenship (SKMF). The movement had won, but victory did not signify that men and women were equal as citizens. Success led to decline of the women's movement. It laid in abeyance for decades. Movement organizations survived without much support and began once again to concentrate on the issues of women's employment and education. Women's caucuses associated with the political parties formed and mobilized politically-interested and newly-enfranchised women.

These two examples show how democratization, a reform so central to and highly valued by social movements, led to their decline as politically viable organizations. Attainment of social movement goals frequently leads to organizational decline. Collective action organizations find that they cannot mobilize support for less central policy issues, and members lack the sense of urgency for a cause that has already been won. Decline took different forms. All Swedish social movements– with the notable exception of unions– experienced a brain drain as electoral reform made political parties attractive for their activists. Parliament replaced public areas and streets as the arena for politics, and collective action organizations were forced to adapt themselves to this new political reality. The emerging political parties began to replace social movements as agents at the cutting edge of change. They became concerned with organizational maintenance, government patronage, and most importantly goal attainment through legislation. Social movements became established and legitimized and began to lose their social movement character. As discussed in chapter 8, some of them were considered so powerful that parliament debated their status and government control of them. The general issue was their role in political democracy.

Labor conflict and prohibition

The turn of the century was a time of political tug of war in Sweden. The new political actors from the formation period competed and cooperated for influence and power, and civil society gave birth to newer collective action organizations. The most significant events of the period are the struggle for universal suffrage, general strike of 1909, and referendum on prohibition in 1922. The general strike was a struggle between organized workers and a new collective action organization, the organized employers. Other social movements were also involved. The referendum on prohibition was one of the greatest mobilizing events of Swedish civil society.

Studies in history have traced the events leading up to the general strike of 1909. Striking was the main tactic used by unions to attain their goals. The political strike of 1902 made the labor movement confident of the effectiveness of striking and also led to the establishment of the main employers' association, the Swedish Employers' Association (SAF), which a rather weak organization with few members and without much opinion formation potential. Other associations and peak organizations for employers were also founded as a reaction to the political strike. Employers organized in the same sequence as workers: first out were the employers for the skilled trades. The employer associations for the skilled trades were initially more positive to collective action and bargaining than the associations for factory workers, which began appearing after 1902.

By 1907 Sweden was an industrialized nation, and unionization advanced at a rapid pace. Union membership rose 260 percent between 1902 and 1907; newer groups– unskilled factory, transport, forestry, and agricultural workers– accounted for a large proportion of this increase. Women represented almost 10 percent of the membership. These were boom years for the employers' associations as well, and the SAF accumulated new and strengthened its older organizational resources (Åmark, 1986, Schiller, 1967).

The Swedish system of collective bargaining began to develop. The first direct negotiating session between the LO and SAF led to an important agreement in 1906, the 'December Compromise'. A crucial clause (paragraph 23) in the agreement dealt with employer prerogative on employee hiring and firing. Other collective agreements were signed. Labor market organizations recognized, with few exceptions, that regulated negotiating procedures were better than government intervention or labor unrest. This meant that employers acknowledged the right of unionization.

The years of labor market cooperation and organization and of economic growth were followed by an economic depression (1908-9), labor unrest, and a new attitude on the part of the SAF. It modernized its organization, centralized authority to the top of its hierarchy, and engaged in labor market conflict to attain national collective agreements. New member associations were established as well. There was tension in the air. Strikebreakers even formed an association as a precaution against upcoming labor confrontations. At the same time, the labor movement experienced internal crises, and a number of local unions left the Social Democratic Party in protest. Even the LO lost members, and reorganization was a hot agenda item scheduled for its upcoming congress. Many employers and union members wanted a showdown in the labor market: a faction within the SAF wanted to crush the union movement and weaken the relationship between the LO and Social Democratic Party; a faction within the LO, which later broke off and formed and an anarchist union (the Central Organization of Swedish Workers (SAC), argued that the reformist line taken by the labor movement was futile. A general strike was the alternative favored by most participants, even reformist social democrats. Local unions were eager to match strength with employer associations, and the LO did not have enough central authority to ward off conflict.

When the SAF proposed a collective settlement to a number of disputes and threatened the LO with a general lockout if it refused to accept it, the LO declared a general strike. Experience with general strikes in other countries encouraged LO's leaders to act in this manner. The business-oriented, conservative government followed the desires of the employers and did not intervene in the strike. Roughly 300,000 workers were involved.

Swedish public opinion grew progressively more negative. A government mediator was able to reach a settlement for certain labor market organizations, but the two main combatants, the LO and SAF, refused to bargain with each other. The strike lasted until the LO was forced to call it off almost five weeks later. It refused to accept the proposal submitted by the SAF in its ultimatum, and a lockout of its members continued, though workers who left the LO were allowed to return to work. The lockout remained formally in force for over a year in some industries (Schiller, 1975).

The general strike is one of the largest labor market conflicts in Swedish history. It had devastating effects on the LO, whose strike fund was depleted completely. Also, the LO lost almost half of its members, and it took almost a decade to remobilize strength. The most active participants in the strike were arrested, forced to pay fines, or blacklisted by employers. Many strikers chose to emigrate. The SAF was the victor of this labor war, and the LO found out that its strength was very dependent on the state of the Swedish economy. A labor leader summarized his experiences from the strike quite cogently:

> Swedish workers had never before experienced any real defeats by their union organizations. They considered setbacks as a step on the road towards success. That Sweden could experience an economic slump which caused the center of power to move closer to the employers was a new idea for the majority of workers. They did not want to consider it because they would be forced to acknowledge that organizational relationships could be problematic. It was these relationships which needed to be reconstructed in order to guarantee sufficient strength to make it possible for the workers, even in depressions, to meet with the same success as they have in times of economic growth (quoted in Hansson, 1936, p. 76, my translation).

Experience from the general strike was not only negative. It showed that the working class was disciplined and steadfast, a most important trait in Sweden. Worker sobriety and the temporary prohibition imposed during the strike heartened the teetotaling movement. Also, the strike showed that the union movement had instilled workers with a sense of self-pride and willingness to fight for their rights against all odds. These rights were not only economic in nature. They concerned, as stated in a workers' newspaper, 'larger demands on life and thereby larger demands on oneself' (quoted in Gårdlund, 1942, p. 427, my translation).

The strike had significant impact on labor market organizations. The LO became more reform-oriented, a development which forced the radical wing to leave in 1910 and form the SAC. Rather surprisingly, organized business experienced a setback. The outbreak of World War I (1914-18) explains this negative development. Labor market peace reigned during the first war years, but the famine of 1917 heightened the class struggle, strained relations between employers and workers, and led to internal problems for the labor movement.

Many organizational changes occurred, and legislation to tame the unions was discussed in parliament. Anti-union sentiments were checked by the suffrage reform of 1909, which enfranchised virtually all men and led to electoral successes for the social democrats. The year 1909 was, without a doubt, one of the busiest and most notable in Swedish political history. Workers were either on strike or locked out; parliament reformed the electoral system, and the teetotaling movement mobilized support for political confrontation on the issue of prohibition.

An umbrella organization for the teetotaling movement was founded in 1908 to mobilize support for prohibition. Its efforts proved successful. A survey of public attitude showed that 99 percent of participants (representing 56 percent of the adult Swedish population) were for prohibition (Lundkvist, 1968, 166). The result forced the government to commission an investigation on the matter in 1911. The IOGT President chaired it, and many of its members came from this and other social movements. The Social Democratic Party voted at its congress in 1911 to include prohibition in its platform, though party leader Hjalmar Branting was a staunch opponent. Mobilization efforts led to increases in membership in the teetotaling movement. By 1910, it had 450,000 members. An official conference of the teetotaling movement took a clear stand for prohibition in 1911.

The situation looked promising indeed. Influential politicians favored prohibition. Electoral reform worked in the movement's favor; after the 1911 election, a sizeable part of the Lower Chamber (59 percent) and not an insignificant part of the Upper Chamber (26 percent) were teetotalers. The proportion rose for roughly ten years, continued to remained high, and then dropped after the national referendum in 1922. Many MPs were members of the IOGT (Meijer, 1956, Johansson, 1947).

The first real movement disappointment came in 1917, when an important goal of the movement, the local veto, was watered down in a compromise in parliament and a rationing book system was applied nationally. The local veto would have legalized prohibition in different parts of the country, after local inhabitants were consulted in local referendums. The movement saw these measures as failures because the goal of prohibition could not be compromised. Members criticized the rather moderate strategy adopted by movement leaders and compared their failure with movement successes in Iceland, Norway, Finland, and the U.S. With these countries as role models, the Swedish teetotaling movement stood more united on the issue and mobilized all its strength for its enactment. Failure on the issue of local vetoes led to a heightened sense of urgency for a national law on prohibition.

Only a general account of the referendum is given here, as the subject is not well-researched. Movement network campaign organizations were created. Free churchmen and -women from all faiths, but particularly the Baptists, who had worked with the issue in both a practical and political way for years, decided to establish a campaign association for themselves in 1920. In 1921 it became a member of a cross-movement campaign organization that also involved the umbrella organization mentioned above. Social democrats formed in 1916 their own teetotaling group. At the initiative of the White Ribbon Society, women involved in the teetotaling and free church movements organized in a special network just for women. Opponents to prohibition also mobilized themselves in 1922. Their organization was called the National Association for Sobriety Without Prohibition, an interesting name for people wanting to imbibe spirits.

Opinion formation was the foremost tactic, and a poster war broke out during the campaign. Proponents passed out almost twelve million campaign brochures and over one and a half million handbills. They set up over 250,000 agitation posters, sent over 400,000 agitation letters, used about 330,000 agitation stamps, and flew 20,000 demonstration flags. Ten thousand speeches were given, and the campaign newspaper with 60,000 subscribers was distributed in 1,300,000 copies (Lindorm, 1958, 143). The figures show the mobilizing power of collective action organizations. Hundreds of thousands of supporters were mobilized to finance and staff the operation. (The population of Sweden in 1922 was roughly six million.) Opponents also actively used public relations tactics, and they had one of Sweden's best political cartoonists in their camp.

It was Sweden's first official referendum, a constitutional change legalizing referendums was passed in 1922. Jubilation over the referendum was muted for some collective action organizations when parliament decided that women's and men's votes would be counted separately. Women's organizations were outraged. A women's demonstration outside parliament had to be dispersed by the police; there were threats of sabotage on election day. Proponents of segregated voting argued that it was the only way to determine whether prohibition could be implemented successfully. The implication was that a majority for prohibition overwhelmingly based on women voters should not be allowed to dominate a majority of men who voted against it.

Not all collective action organizations accepted this reasoning, but the IOGT leadership did with the motivation that the referendum was more important than equal voting rights. The principle did not need to be applied, but the vote was close: 51 percent against and 49 percent for prohibition (Wallin, 1966).

After all its effort, the teetotaling movement was stunned by these results. It had lost this most crucial test of public and political strength. Movement organizations evaluated the defeat differently. The IOGT was not yet ready to give up the fight as long as prohibition was still law in the U.S., but it began to devote its attention to other activities, in particular internal organizational matters. The NTO blamed the defeat on the press, which it considered the 'motor of the opposition', and the better economic position of the campaign against prohibition (Svensson, 1979, p. 236, my translation). Negative experience with prohibition in the U.S. gradually changed the IOGT's position.

Maintaining social movements organizationally

Social movements went through a number of trials. They became legitimized public actors as well. This public status implied that the political establishment viewed them in a positive light. No longer was their behavior considered to be politically deviant; nor were they considered as pestering protest groups. Their leaders were people consulted by government. They had gone through the first phase in their life cycle and were now institutionalized.

Several social movements began to receive government patronage. Frequently a sympathetic MP motioned in parliament for funding for collective action organizations. They eagerly accepted monetary support not only as a gesture of appreciation but on economic grounds. Collective action organizations experienced financial difficulties, which membership fees could not redress. Their buildings and meeting halls were in need of repair, and they found organizational maintenance a costly enterprise. The general strain of World War I and the economic recessions of this period affected the social movements as they had most of Sweden. Many members left movement organizations because they could not pay their membership dues.

Economic problems were closely tied to other problems. Many collective action organizations found it difficult to recruit members and were losing members. Decline was temporary for some movements and permanent for others. Union membership stagnated after the general strike until 1911, and membership in the women's and peace movements never really stabilized. Membership in the teetotaling movement peaked around 1910 and for the classic free churches in the 1920s. There are several reasons for recruitment problems.

First, members were now less committed. Membership was no longer a life style for many people, as other matters, interests, and newer collective action organizations offered them other alternatives. The thrill of belonging in the genesis phase was now replaced by a more realistic assessment of the potential of membership. Adjustments were made to arrest the exit of members, but some of them only alienated members further. Second, many people joined movements in the past for their entertainment activities, a type of selective incentive at the time. For instance, people took the vow of abstinence just to be able to attend dances sponsored by the teetotaling societies, but now they could entertain themselves without becoming members. The availability of other sources of entertainment meant that collective action organizations lost their monopoly on leisure time activities. The result was fewer but more dedicated members, though this did not necessarily mean that they were actively involved in movement affairs. Third, traditional leadership, which instilled members with enthusiasm, was replaced by leaders more concerned with statesmanly behavior and organizational legitimacy. Administrative skills were emphasized as important leadership qualities. Many capable leaders even left social movements for political parties and distanced themselves from movement principles. An important exception here is the LO. Fourth, the practical difficulties of maintaining collective action organizations in the war crisis taxed members and leaders psychologically and physically. Many of them had overcommitted themselves. Movement failures and disappointments did not help keep the spirit up. The general strike and referendum on prohibition offer the clearest examples of the effects of expectations-gone-wrong.

After the strike, the LO lost much of its authority over its unions. Members left in droves. Unions organizing women, weaker, and less attractive groups and unskilled workers suffered most; many of their local branches closed down. Member activity was generally low, and many activists left the movement. Organized employers took advantage of the situation in the collective agreements signed in these years. Defeat in the referendum on prohibition had a similar effect on the teetotaling movement. Pessimism replaced enthusiasm and optimism; many members lost faith in the cause. Declining membership rates and financial difficulties could not be dismissed. The teetotaling movement was forced to acknowledge that it no longer channeled the needs of large groups of people. Different approaches were tried to improve the situation, but the most important key to the future was government patronage. New goals were added and selective incentives used to recruit members. Feature films were shown to entice people back into meeting halls. More traditional movement sympathizers considered all departures from traditional goals and tactics as demeaning for the cause and leading to the depravation of the youth. It is noteworthy that the only teetotaling organization not to decline was the White Ribbon Society. This development is attributed to its broader goals, which included not only abstinence but women's, peace, and social issues. It devoted considerable energy to help unfortunate women improve their life chances.

Other collective action organizations experienced problems as well. The peace movement was frustrated. Its basic problem was that it could not mobilize supporters over time to become institutionalized. It experienced its third and fourth mobilization peaks in this period. People flocked to the SFSF before and during World War I, but many of them left once it ended. Several small peace organizations were created but could not gain sufficient public support for survival. Recruitment efforts by a charismatic movement entrepreneur explain much of the upswing in members for the SFSF in the 1920s. But, once again, the movement had difficulty in holding its members, and membership dropped in the 1930s.

A vexing problem for many organizations was recruitment of young people. The vigor of youth, which characterized their genesis phase, was now replaced by middle-aged complacency. Part of the problem was that there were more opportunities open to young people. They could choose among the classic social movements or other collective action organizations emerging at this time. The sports' movement was particularly attractive, as was commercial entertainment. The other, more serious development was gradual detachment of young people from the spirit of the collective action organizations. Many organizations were forced to give activities for young people lower priority during World War I, with lower membership rates as a result. Also, they continued to use the same (now old-fashion) recruitment methods, which did not lead to positive results. A controversial, new suggestion was the establishment of scout organizations. Many movements established separate youth associations. Sympathizers with foresight realized that a collective action organization that could not renew its membership would stagnate and die eventually. The first important new recruiting ground was the youth, in particular boys and young men.

The only movement that did not degenerate was the LO. It grew strong after recovering from the setback of the general strike. This is partly explained by its general appeal and ability to attract adherents once society underwent change. It could keep its sense of urgency at a high level. By the end of this period, the union movement succeeded but the free church, peace, women's, and teetotaling movements failed to sustain this crucial kind of support mobilization. Success here is related to the piecemeal achievement of its basic goals: members were kept at urgency alert every time a collective agreement was bargained or strike or lockout called. Class-based politics also kept the sense of urgency high.

The second explanation concerns the goals of collective action organizations and membership incentives. The unions combined both material and non-material elements in its objectives, while other movements generally focused only on non-material goals. This combination gave them a stronger calculus of group membership, which appealed to individuals whose entrance and retention motivations varied over time and among member groups. Also, selective incentives were in operation: workers received strike compensation, and the unions financed and managed the unemployment societies.

Finally, the LO did not lose its social movement entrepreneurs to political parties. The union movement remained dominant even though it was more closely associated with a political party than any other collective action organization. Perhaps the association with one particular party proved beneficial. The union movement was able to recruit top leaders from its own ranks, which continually expanded over time. There was a remarkable stability and continuity in the LO leadership when compared to the other movements, whose leaders seemed to follow a sense of urgency which took them from one collective action organization to another. It was possible– even in this trying period– for a gifted member to make the union his career. Later this ability would lead to oligarchical tendencies. Finally, the class-based character of Swedish politics helped the union maintain its members and public status.

Other social movements join Swedish civil society

Collective action organizations belonging to the second wave of organizational development emerged and gained strength. Their cause was economic rights, and they fought for more equal distribution of society's economic resources. Their mobilizing successes led other societal groups to react, and they formed their own organizations to counteract the presence and growing strength of existing groups. The SAF and other employer associations are good examples of countervailing powers. Social movements– many of which were third wave organizations– emerged to meet the needs created by the changes society went through in this period. Their goal was citizen empowerment. Finally, disagreements within the first social movements led to splinter groups, which broke off and formed their own organizations.

As with all other movements, the origin of the Swedish consumer cooperative movement was international and based in the changes underwent by society. The roots of the movement are from the early 1800s, but movement growth was inhibited by a law prohibiting cooperatives that was repealed in 1846. The same social changes that explain the emergence of the union movement apply to the consumer as well as farmer cooperatives. A particularly important development was the emergence of organized capitalist economy (Torstendal, 1991).

Sweden's first consumer cooperative societies were closely allied with the liberal union movement. An unsuccessful attempt was made in the late 1800s to implant a Belgian model that recruited exclusively among union members. The cooperative movement really began to mobilize support in the late 1890s, when the Swedish Cooperative Union and Wholesale Society (KF) (founded in 1899) brought together geographically dispersed consumer groups into a national, peak organization. It welcomed all social classes as members. Other consumer cooperative associations as well as the Women's Consumer Guild Union (KK) were founded as well.

There was strong initial reaction against the KF. Banks and the food industry boycotted it, and an association for retail business was established in 1908 to counteract it. Their behavior is understandable, for consumer cooperatives were formed as reactions against capitalism, monopolies, and middlemen. Business and organized consumers fought for control of the retail market. Initially the prospects looked very bleak. Public support was weak, and its relations with the labor movement were not good, though many social democrats were founding members. Party leaders were skeptical about some of the movement's principles, in particular its nonpartisanship. Also, members had a very vague understanding of the goals of the movement, and its organizational structure left much to be desired. This situation characterizes most social movements in their genesis stage.

The movement was resilient and benefitted from the attacks. Public interest was aroused, and sympathizers mobilized to support principles of consumerism. Members and leaders began to realize that a strong organizational economy and structure were important and that opinion formation was the best way to reach out to the public. A newspaper and school for educating leaders were started. The economic situation of the war years helped change the attitude of Swedes to the movement, and a large number of local consumer societies were founded. The KF underwent reorganization and professionalization. It purchased companies in the food industry to improve the quality of products for its members and force competitors to adjust market prices. By the end of this period, the KF was a strong organization, and consumer cooperatives well-received in society.

The cooperative idea found fertile ground in Sweden, as witnessed by the founding of the housing cooperative, the HSB, in 1924, other consumer cooperative organizations and, of course, the farmers' cooperative movement. Appendix 6 offers information on membership in the consumers' and housing cooperative movements. The history of farmers' cooperatives is different than the other cooperative movements. Swedish government took an active part in their operation from the start, a development explained by the importance of a prospering agricultural sector for national defense. Dairy, egg, and meat cooperatives appeared in the late 1800s. An important event was the passage of a law in 1895 that created a new legal category, the economic association or cooperative. It helped the existing cooperatives stabilize their financial situation and benefitted farmer as well as consumer cooperative endeavors.

The first national farmers' cooperative association, the Swedish Farmers' Supply and Crop Marketing Association (SLR), played a major role in helping government distribute food in World War I, and the legitimacy it received for its efforts led to success in member recruitment. A national association, the Swedish Agricultural Society (SAL) brought together individual farmers and different cooperative organizations. Its aim was to establish farmers' cooperatives in new fields and persuade existing cooperatives to become members. The SAL did not prove to be an effective association, and its membership shrunk as the agricultural crisis of the 1920s worsened.

A fourth collective action organization to attract supporters was the sports' movement. As with many others discussed here, it dates back to the 1800s, with the 1880s as a particularly important decade. Young people found the movement highly appealing, though it was initially closed to women. Its presence meant that some of the older collective action organizations now had to compete over members, which led them to debate whether to establish some kind of sports activity. The teetotaling movement was the only one to do so.

A national association for gymnastics and sports was founded in 1903. Industrialization and urbanization were the societal developments that spurred its growth. Young people living in towns and cities found the movement most appealing. It emphasized the importance of sports for patriotism and Swedish defense– the two hot topics of the day– and was closely allied with government from the start. The association received its first government grant in 1913; it was a sizeable sum when compared to the patronage given to other collective action organizations. The political left was very skeptical of the movement; the political right was very positive. The left called it militaristic, bourgeois, and royalist and questioned whether it could rightly be considered a social movement for it lacked ideals and disregarded cultural politics (Lindroth, 1974). It is worth emphasizing that sports' associations are classified as a social movement in Sweden and receive tremendous amounts of money from government. Appendix 7 offers membership statistics.

The changes that Sweden underwent since the mid-1800s are symbolized in a poignant way in the emergence of two final collective action movements, the adult education and conservation movements. Popular education became a key activity within the first social movements in their genesis stage, and the need to educate members took organizational form in study circles early in their institutional stage. Though imported from abroad, study circles found very fertile soil in Swedish civil society. They became a natural vehicle for organizations to advance their causes and create educational opportunities for members with limited access to formal education. After the government lessened its control over this type of activity in 1912, adult education societies were established as coordinating units for the study circles. One of the first was the Workers' Educational Association (ABF). Seven years after its establishment, the ABF applied for and received government patronage. Gradually every social movement established its own adult education association. The political democratization of Sweden and enactment of the eight-hour day are two factors that encouraged the growth of this movement.

Not everyone viewed industrialization and urbanization as progress. A group of concerned citizens– the conservationists– attempted to mobilize support for what they saw as the vandalization of nature. Conservation was an international movement at the turn of the century. National conservation societies had already emerged in several countries. Swedish conservationists began to mobilize support for a peak organization in 1908. They met resistance from representatives from existing organizations who argued that another organization was unnecessary.

It was initially difficult to agree on organizational goals. Internal disagreement would continue after the Swedish Society for the Conservation of Nature (SNF) was founded in 1909 and hamper its recruitment efforts. The SNF began as a revival movement of sorts for it aimed at 'awakening love for Swedish nature' (Haraldsson, 1987, p. 18, my translation). It attempted to broaden its base through membership campaigns and by specifically employing people to mobilize local support. It is of interest that the SNF worked closely with societies for artists, writers, and young people, i.e., people having 'access to a number of immaterial resources, such as knowledge, prestige, and different channels to relay their message' (Haraldsson, 1987, p. 68, my translation). This shows that the movement gave opinion formation high priority.

All attempts failed due to lack of necessary organizational resources, and the SNF was forced to abandon its dream of becoming a true social movement. Instead, it developed into an organization of and for experts. Many of its founding members were scientists and radical intellectuals; its first board was composed of men– no women– from higher echelons of society, and its structure was hierarchical and centralized. It received financial support from government from the start.

There was a clear link between the SNF and local cultural societies, collective action organizations for preserving yeoman culture, and the sports' movement. Other organizations were involved in this network, and the SNF took the initiative to organize them to put pressure on government and influence public opinion. One of its main activities was lobbying government to enact laws to protect Swedish forests, air, undammed rivers, and cultural landscape from exploitation by industry. The conservation movement was not entirely unsuccessful in its efforts.

Its center was the cities of Stockholm and Uppsala. It emphasized the importance of recruiting young people, but its image prevented any large mobilization of support. Membership rates were unstable and very modest (appendix 8). Decline was rather rapid after 1920; this led to reorganizational efforts. Professionalization and specialization, better public relations, and selective incentives to entice people to become members were suggested strategies. Yet the root of every SNF problem was disagreement over its basic goals.

Changes took place also within the classic social movements. Many of them expanded their scope of activity. For instance, an automobile insurance club for teetotalers (MHF) was established in 1926, and the labor movement began to construct people's parks, which offered sympathizers a place of entertainment and impregnated them with movement values. Other changes found their origin in irreconcilable differences within the classic movements, and a number of followers were forced to form their own organizations. Two of the best examples show how established Swedish political actors attempt to quench the emergence of competing new collective action organizations.

A radical syndicalist, communist faction within the LO broke off and formed the SAC, which organized both blue- and white-collar workers from all employment sectors. Syndicalist unions were created in many countries. Oppositional groups within the LO falsely believed that the SAC would be able to mobilize considerable support. It was shunned by the social democratic labor movement, government, and SAF, which refused to bargain with it. That it scorned collective bargaining and believed in striking did not improve its image among political elites. The SAC was unsuccessful in its few attempts to seek cooperation with the LO in the early 1900s. Yet several of the ideas raised in LO debates find their origin in the SAC (see Persson, 1975). Relations between the LO and SAC have been strained throughout this century.

The second example is the Pentecostal movement, which was brought to Sweden from the U.S. by a Swedish preacher in 1907. It attracted good leaders and appealed to the Swedish sense of pietism. Workers and peasants– not the middle class– were drawn to it. Most members were rather young, which is of interest as free churches were experiencing problems with youth recruitment. The Pentecostalists began as a protest wing within the free churches but soon became isolated and were forced to break out and form their own church. Their religious beliefs were seen as a threat to society and Swedish Christianity because they were seen as primitive and foreign. Government viewed the movement with suspicion; the state church considered it a 'dangerous delusion' (Thulin, 1945, my translation). Pentecostal leaders were questioned by the police, and the classic free churches thought it a menace and misfit. Most Pentecostal followers had their religious roots in Baptism, which explains the particularly harsh criticism by Baptist churchmen and -women. Even the mass media lashed out at it. The press interviewed experts who condemned the religious belief on the grounds that it led to psychological and physical disorders (Coleman, 1991). It was even castigated for being too American in orientation, a rather amusing criticism as almost every Swedish free church had its origin either in U.S. or English religious revivalism. In sum, the establishment used all available arguments to label it as dangerous and undesirable.

Followers of Pentecostalism denied these accusations. They saw their church as returning to Christian origins. Sympathizers criticized the bureaucracy, alienation, and spiritual laziness which they believed characterized the established free churches. They even viewed the teetotaling and peace movements disapprovingly, for they were seen as focusing on simple matters and not aiming to transform the entire human being. Founding members of the free churches reasoned in a similar way about the state church in the 1800s. There are obvious parallels between the emergence of the first free churches and the Pentecostal movement. The same criticism had been directed at the first free churches, which shows that the Swedish establishment of the early 1900s chose to recall only those aspects of history which furthered and protected its own interests. This illustrates a behavioral characteristic common in many cultures, i.e., historical forgetting and the conscious formation of collective memories. This would later would lead to the glorification of first and second wave organizations in Sweden.

4 Creating the 'strong society'

The boundaries of the Swedish political began to change radically and would continue to change in this period dated to the 1930s. Swedish politics had became class politics from this period and dealt with problems of material welfare (see Therborn, 1992). Every social class had a political party to represent it. The main political actors were the political parties and labor market organizations. This meant that demands of other collective action organizations were either given lower priority or accommodated to the goal structures of these main actors. Compromise characterized this period, and there were few open struggles over political power. The country was tired of conflict. War crisis instilled as sense of nationalism and social responsibility. Defending and building up the nation was the foremost priority, and political architects came from the political parties LO, SAF, and farmers' movement. This was the basis of the strong society.

One of Sweden's most prominent politicians, Tage Erlander, coined the term the 'strong society' in the 1960s to capture his and social democratic expectations (see Tilton, 1992). These hopes involved future opportunities for young people and development of a welfare state. The term 'strong society' fits this period. It summarizes its main social, economic, and political developments and the role that collective action organizations played in them.

Social movements were still important, but government became the main force that reshaped society and changed the boundary of the political. The government, parliament, and expanding public administration were now the cutting edge of politics. Swedes came to believe that political change could only occur through government action and that the only truly effective collective action was taken by government. The term 'strong society' also symbolizes the demise of the independent civil society that grew strong in the second period. The clear distinction between civil society and the state disappeared from the Swedish mentality. Government and civil society became embedded in each other. Few people viewed this development critically, for the strong society linked government with civil society and, under the specific circumstances of the period, accomplished many impressive, and, for other countries, enviable political tasks. Contemporary Swedish dictionaries and lexicons do not even distinguish between the words 'society' and 'government'.

The Swedish folk, nationalism, and social movements

Social movements of the formative period were significant agents in furthering respect among social classes. Respect for class differences became embodied in the concept of the Swedish people, an organic concept stressing the collectivity rather than the individual as the basic unit of society. The concept of the people paved the way for the development of the universal welfare state in Sweden. This is a somewhat unique development, as shown by a comparison with Germany, where the concept of the people led to anti-democratic and racist sentiment and later national socialism (Trägårdh, 1990).

That Swedes perceived their nation as homogeneous– i.e., without serious ethnic and religious dissension– and free from threats from foreign powers made it easier for this type of democratic, collective identification to develop. The European political situation helped form this identity by forcing a sense of national solidarity on people. It made them feel that 'all members of society are in the same small boat, that the waves are high, and that everyone must help pull the oars'(Katzenstein, 1985, p. 35). Also, the reformist orientation of Swedish social democracy stressed cooperation, pragmatism, and compromise rather than conflict with and impertinence towards the capitalist class. One of the best examples of this national mentality is the Agreement of Saltsjöbaden, as discussed later.

The Swedish concept of the people implies social solidarity and responsibility. Its origin as a concept for equality among social groups can be traced far back in Swedish history. Social democrats adopted the concept as their own at a very early date, but it was not given higher priority than the concept of worker until this period. Competing interpretations of the concept existed, as illustrated in the origin of the term 'people's home', which was coined by one of the foremost opponents of universal suffrage to symbolize the organic conception of the conservative, corporate state.

The term 'people's home' was then captured by the social democrats and later became a Swedish trademark. The leader of the Social Democratic Party, Per Albin Hansson, used it in a parliamentary debate in 1928 to symbolize responsible, democratic governance. Afterwards, he declared:

> the concept of people can open the road to cooperation in those areas where the concept of class seems limited and isolated; while the concept of people focuses attention more on using resources of power cooperatively so that misery can be alleviated, the concept of class can easily lead to the avoidance of responsibility (quoted in Tingsten, 1967, p. 295, my translation).

This concept of people became firmly rooted in Swedish mentality. The Social Democratic Party successfully used it in the 1936 parliamentary election campaign, and the LO acknowledged that development of the people's home was more important than class conflict. Acceptance of the term allowed for the integration of political democracy, the people, social democracy, and social movements. This integration is the foundation of modern Swedish political culture.

New power structure

Universal suffrage transformed Sweden's political landscape in several ways. It led the liberals to withdraw from the left coalition and allowed the social democrats to take office for a very short time in 1920. Weak minority governments characterized Sweden for over a decade. Instability typified Swedish society. Few collective action organizations understood the implications of minority parliamentarianism for the political opportunity structure of Sweden. There was labor unrest, and the economy only seemed to get worse. The LO became a battleground for communist and social democratic members at the same time it was trying to recuperate from a scandal that forced its president to resign. Union wage policy was blamed for the high rates of unemployment, and government contemplated anti-union legislation.

In 1928, the government created an organizational setting for cross-class discussions when it called the SAF and LO to a meeting to deal with the problem of labor unrest and legislation that encouraged labor mediation and regulation rather than conflict. But by 1931, it became clear that class conflict crushed any hope of cooperation. The disturbances in Ådalen, an area in middle Sweden hit hard by unemployment where four demonstrators and one bystander were killed by military bullets, symbolized the inability to reach agreement across social classes.

Sweden was in desperate need of political stability, and cross-class compromise was the way that it was achieved. A 'crisis agreement' was reached by the social democratic minority government and Agrarian Party in 1933. Three years later, the parties formed a coalition government and ruled jointly until the outbreak of World War II. The background of this 'historic agreement', as it has been labeled (see Korpi, 1978), is well-documented and will not be recounted here (see Söderpalm, 1975, Lewin, 1975). Suffice is to say that political actors who would be decisive for decades to come had now made their entrance in established Swedish politics. They are the political parties, social democratic government, LO, farmers' organizations (RLF and SL), big business, and the SAF. They represented the networks of organized workers, farmers, and business. The crisis agreement of 1933 was a prerequisite for the social democratic road to power.

The social democratic government benefitted from the improvement of Sweden's economic situation that began in 1933. Better economic conditions allowed it to begin to realize its goals of the people's home. Social policy legislation included government-sponsored unemployment insurance, general folk pension, and eight-hour workday for farm laborers. Tax policy changed in order to finance welfare reforms and reduce differences among the social classes. This was the public policy of the people's home. The social democrats formulated a new economic policy which, in international circles, became known as Sweden's middle way (Child, 1937). It involved increased government control over economic life without revolution or large-scale nationalization. Sweden's reputation as a high-tax country began in this period.

The important implication of this development is that the social democrats no longer assumed that an independent civil society was an essential safeguard of democracy and cutting edge of politics. Rather, free elections were assumed to ensure political development along democratic lines. Social movements were seen as part of the political order, and this affected the development of collective action organizations in several important ways.

Accepting public responsibility

For many social movements, the crisis agreement of 1933 meant that their political claims would be overshadowed by the prerogatives of organized business, farmers, and workers. To gain political influence they would not only be forced to appeal to government but to the interest organizations of capital, labor, and agriculture. Also, the new power structure of Swedish politics had significant impact on labor relations. The best example here is the Agreement of Saltsjöbaden between the LO and SAF in 1938. It is a favorite subject of labor market historians and scholars of the Swedish model. They write with pride about a special Saltsjöbad epoch and spirit. The agreement came to symbolize the Swedish political culture of compromise across social class boundaries. Yet if looked at closely, it was, in fact, the beginning of a labor market cold war that lasted until the mid-1970s.

The background of the agreement was a balance of power between the LO and SAF. The LO had by the mid-1920s recovered from the ruinous drop in members from the time of the general strike, and the labor movement was mobilizing electoral support. Also, the SAF was gaining in strength. Employers had built up a viable labor market association and were now turning it into a very effective counterpart to the LO. Choice of organizational weapons was debated lively within the SAF.

Government generally feared a labor war, the outbreak of which would devastate Sweden and distort the public agenda for years to come. It pondered the dilemma of democracy– i.e., how it should balance the need for control of the labor market with the democratic right of employers and workers to freely join associations. The organizational setting established by the government in 1928 had failed to improve the situation, and strikes and lockouts were on the increase by the early 1930s. A government commission, which was appointed in 1934 to investigate problems in the labor market, proposed legislative action as well as democratic and structural changes within the LO as a possible means to achieve labor peace. Yet before such action was taken it believed that the government should sound out the labor market organizations on their willingness to reach labor peace on their own.

This occurred in 1936. The SAF reacted strongly against the government's attempt to gain control over labor market organizations and decided that it was better to negotiate directly with the LO. This put the LO in a delicate position, but it decided to negotiate with the SAF because an alliance with the government was too risky. Neither the LO nor the SAF could count on a dominant political alliance in parliament. Not all social democratic cabinet ministers were happy with this solution because it meant that the government would not be a welcomed participant in Swedish labor market relations. The decision by the LO and SAF formed the basis for relations between them for decades to come. Their decision is another example of the impact that Sweden's political opportunity structure has on collective action organizations.

The SAF and LO wasted no time and began to negotiate in 1936. An agreement was signed in 1938. Soon afterward the agreement gained mythical proportions. Yet when examined closely, the basic contractual document is 'elusive and not very obligatory' and simply reaffirmed the collective bargaining traditions already practiced by many of the larger LO and SAF member organizations (Alf Johansson, 1989, p. 71, my translation). It did not centralize collective bargaining to the SAF and LO but only obliged them to seek to make their member organizations understand the importance of labor peace. Neither did it apply to the entire labor market: the construction industry, public sector, and white-collar workers were not included in the agreement. The significance of the agreement in and outside Sweden has led labor market historians to ponder whether a more substantial agreement had not been secretly reached by the LO and SAF (Edlund et al., 1989). The left-radical faction within the LO criticized the LO leadership for selling out union democracy for they saw the agreement as leading to union oligarchy. Over the next ten years, the spirit of Saltsjöbaden led to other cooperative agreements between the two main peak labor market organizations. The number and extent of labor conflicts decreased drastically, but it cannot be known whether the economic crisis of the 1930s, World War II, or spirit of Saltsjöbaden deserve most credit here. Yet sheer reference to the Agreement of Saltsjöbaden forced labor market organizations to reflect carefully about the pros and cons of labor conflict.

Most writers have viewed the Agreement as a goal in itself. They argue that the general objective was labor peace and cooperative arrangements between the central labor market organizations. This is not true. Labor peace was the goal of the government, but it was more a means than a general goal for the labor market organizations. The LO and SAF agreed on the need for structural rationalization of industry, economic growth, and labor peace, but they did so for different reasons. While organized employers saw these measures as leading to higher profits and a better competitive position internationally, organized workers considered them as the basis for a better standard of living. Thus, the two major combatants in the Swedish labor market found a common means to solve ideologically divergent goals. The means used by them in the past to attain the same goals were strikes and lockouts. The means which they used from the late 1930s to the mid-1970s found their basis in the spirit of Saltsjöbaden and involved corporatist integration.

Corporatism or the integration of government and civil socity was a highly debated topic in this period. A prominent liberal politician argued that collective action organizations should not be permitted to 'develop into petty kings who misuse their power and check government authority on decisive issues' (Bertil Ohlin quoted in Söderpalm, 1976, p. 36, my translation). Many concerned citizens of non-socialist persuasion agreed and began to see these organizations as 'a danger for the freedom of individual action, authority of the state, and legal order' (Heckscher, 1951, pp. 158-59, my translation). Another debate topic was organizational democracy: Parliament debated whether to pass a law guaranteeing citizens the right not to be forced to join a collective action organization. This right, which has been debated in earlier and later periods, is called the right of nonassociation.

Yet while the public debated the dilemma of democracy, corporatist meetings between the government and organized business took place, and different social movements and interest organizations were made members of government boards. The worrisome economic situation of the mid-1930s called for political pragmatism involving the use of established, corporatist channels of communication between government and collective action organizations from earlier periods of Swedish history, particularly World War I. There was even cooperation between big business and the social democrats, who regarded the large, monopolistic, export-oriented Swedish companies as important negotiating partners and political actors. Cross-class cooperation proved beneficial for organized business as well as the labor movement. The basis for this cooperative effort was the general economic and political situation as well as the spirit of Saltsjöbaden. A social partnership was thus formed.

It is important not to exaggerate the extent and impact of cross-class cooperation and make it into a political myth. The corporatism that developed did not signify the death of political ideology. The debate on economic planning in the 1940s showed this very clearly (see Lewin, 1967, 1975). There were fundis groups within each base of the new Swedish power structure that disdained compromise and desired more adamant action. Communist union members complained that the LO was selling out the working class; the National Farmers' Union (RLF) threatened a nationwide milk strike during World War II, and the Directors' Club, a group representing large, multinational companies in Swedish manufacturing industry that was created in 1933, wondered whether cross-class collaboration was in its interests. This group believed that they still could influence politics by lobbying government. The problem was that these traditional pluralist methods from earlier periods were less effective in the new power structure of Sweden. Yet cooperation with the social democrats generally led organized business to view involvement in party politics and ideological campaigns in a more negative way.

During World War II, Sweden was governed by a grand coalition composed of the Social Democratic, Liberal, Agrarian, and Conservative Parties. Government reliance on collective action organizations– particularly the LO, SAF, and farmers' organizations– was especially strong in the war years (1939-45). The role played by collective action organizations in World War II has been adequately dealt with by scholars, and their findings will not be repeated here (see Katzenstein, 1985, De Geer, 1991, Micheletti, 1990a, Heckscher, 1951). Suffice is to say that 'no segment of civil society refrained from cooperating with government in the war effort' (Heckscher, 1951, p. 233, my translation). Public responsibility-taking and cross-class cooperation worked well together in these trying times.

Social partnership implied that Sweden was more united than ever before in its modern history. Government was the actor as well as the setting that linked the bases of the new power structure together. Possibly the LO Congress of 1941 best symbolized this linkage. It discussed a report on the union movement and private enterprise. The spirit of Saltsjöbaden as well as the war crisis influenced its contents, which concerned the need to centralize union authority to the LO leadership so that it could discipline union members and take public responsibility. A crucial element in the social partnership was, without a doubt, the centralized authority of peak collective action organizations.

Influential trade unionists believed that the report should be accepted by the congress because public responsibility-taking would not compromise the interests of members. A prominent trade unionist argued: 'No labor leader who assumes public responsibility can deny the obvious fact that individual interest must subordinate itself if asked to do so by the public interest'; another spokesman summed up this sentiment with the words 'we are the state' (quoted in Öberg, 1991, p. 49, my translations). Of course, social democratic electoral successes made it easier for the labor movement to identify itself with government, but the fact that organized business and farmers did likewise shows how deeply collective action organizations were affected by the creation of the strong society.

The message from the LO Congress is of principled importance for this book. Problems that collective action organizations created for political democracy were not unsolvable and did not constitute a true dilemma. The key was participation in political decisionmaking, social partnership, and corporatist integration. It is important to remember that the LO was aware that the public interest could not always determine union actions. It had not forgotten that the primary task of unions was to represent the special interests of its members (see Hall, 1992). The problem was to decide whether member interests were best represented by corporatism or pluralism. The LO chose the corporatist road to power from this period until the early 1990s. The SAF chose the same road until the 1970s. Organized farmers remained faithful adherents until the late 1980s. Government began to reconsider its partnership with encompassing collective action organizations in the 1970s.

Collective action organization maturity

Many collective action organizations matured, some stagnated, and a few laid in abeyance in this period of civil society history. A general indication of organizational maturity was centralization of authority and power to peak units, mergers of local branches into larger ones, transfer of influence from elected leaders to paid staff officials, and social partnership with government. This development applied to the KF and farmers' associations, SNF, and labor market organizations. Many social movements transformed from true protest groups into defenders of the political order.

Different symptoms of organizational stagnation became apparent. It was becoming extremely difficult for some collective action organizations to mobilize internal and external support for their cause. The teetotaling movement experienced serious problems with interest inarticulation and internal unity due to its social partnership with government. It also recognized the need to adapt its goals to current realities. The free church movement was out of touch with social change. Collective action organizations began to reevaluate their emphasis on meeting attendance as a necessary form of movement involvement. They considered the status of passive members and discussed alternative retention incentives to keep non-active members in touch with activities and discourage them from leaving organizations. More contacts among social movements were also proposed, and it was considered particularly important that stagnating movements establish relations with the labor movement.

The Swedish public was not very responsive to gender issues, and the women's movement continued to lay in abeyance. The FBF pressured government with a good degree of success on specific issues related to women and the family, but it failed to mobilize mass support for its main concerns involving equal opportunity for women in working life and representation of women in the public sphere. The concerns did not reflect the zeitgeist: the Swedish people were engrossed with other matters, and the women's issue was overshadowed by concerns of the labor market.

It is not easy to generalize about the situation for the peace movement. Its ability to mobilize support depends always on international events. The analytic difficulty is that external circumstances led both to the rise and decline of members in and successes and failures for the largest movement organization, the SFSF. The issue of peace was part of the zeitgeist, and the SFSF was successful in realizing a number of its goals. The problem was that its conception of the issue did not fit with the desire for national unity that was expressed in this period. Nationalism, neutrality, and military defense concerned the Swedish people, and this explains why the development in Germany in the 1930s did not create a sense of urgency for the cause of peace. The international orientation of the peace movement did not fit with nationalistic feelings embraced by political parties and most collective action organizations. Membership in the movement dropped in the 1930s. The social movement network that evolved around the peace issue in the first and second periods began to fall apart. Network organizations developed perspectives of the peace issue that better suited their own ideological needs. The peace movement became increasingly isolated from other collective action movements, in particular the labor movement by the end of the 1930s.

A general development was that collective action organizations, which were formerly treated as deviant outsiders, began to achieve respectability. Changing attitudes of the Swedish public as well as organizational considerations were the causes. For example, the Pentecostal Movement deemphasized its more deviant characteristics– apocalypse teachings and speaking with tongues– and changed from a religious sect to an established church and mature social movement organization.

It is also of interest that less has been written on the situation of many of the other collective action organizations in this as well as later periods. Numerous books and articles from different dates extol the role of the first social movements in the Swedish democratization process, but they rarely discuss what happened to them afterward. This is how Swedish academia and society has glorified them and molded the Swedish collective memory. We learn about movement achievements in earlier periods but less about their organizational developments and problems in later ones. The exception here is labor market organizations, particularly the LO and SAF. The reason is their extreme importance in most of civil society history.

Labor market organizations were transformed considerably. One major change was the decision to create industrial rather than craft unions. This led to union centralization, amalgamation, and rationalization and eliminated inter-union competition. In the longer run it secured for the LO the role of representative monopoly for the working class. Centralization even implied better discipline of members. Oppositional member groups– communists and anarchists– were silenced. Important areas of responsibilities were turned over to more centralized units. For instance, the right to call strikes was transferred to the LO leadership. The LO's control over members and conflict funds increased greatly after revision of its constitution in 1941, and practices and procedures were standardized. Local units became larger. The SAF also became more centralized as did the Swedish Federation of Industry (IF), a trade association founded in the second period of civil society history.

It was possible to develop standard solutions to problems because interests of individual union members were simplified and aggregated into homogeneous organizational interests. The factory system and technological advancement in industrial production had furthered this development. Uniformity was created through centralization and member discipline (Åmark, 1986). Male industrial workers became the norm for assessing member interest. The special interests of other sectors of employment– service, public, and agricultural sectors– were deemphasized as were the interests of women workers (Qvist, 1974).

Elite democracy governed the unions and other collective action organization (see Micheletti, 1985b). Democratic centralism was the term first used by the LO to characterize its internal democracy but was later replaced by less ideological terms by this period. Members generally accepted the new organizational structures and democratic practices for they benefitted from them economically. The changes furthered the social democratization of Sweden. Most union members considered the presence of the Social Democratic Party in government to be a guarantee that centralized power would be used to further rather than abuse their interests. Centralization, industrialization, material incentives for membership, and the new power structure fit well together. They formed the basis of the Swedish model, as discussed in the next chapter.

Competition over representing interests in civil society

Older mature and stagnating collective action organizations met with competition from newer organizational actors. This development implied vying for individual members and struggles among organizations for political influence. Civil society became populated with more specialized collective action organizations. It was said that all social interests in Sweden were represented by interest organizations or social movements. Swedes have even coined a term for this. It is 'thoroughly organized'.

A number of business associations were founded in these years, many of which had roots in organizational forms from the late 1880s. The SAF now had competition from other associations that organized private employers. They included associations for employers dealing with commerce and trade, banking, insurance, and skilled crafts as well as small industry. A change in the political opportunity structure explains their creation.

New laws granted white-collar workers employed in the private sector the right of association and collective bargaining, and this led to their unionization. Several peak unions for white-collar workers were created in the 1930s. For example, the Salaried Employees' Central Organization (DACO) represented all categories of white-collar workers in the private sector. The Central Organization of Salaried Workers ('old TCO') organized all categories of public sector white-collar workers from 1937 to 1944, when the DACO and 'old TCO' merged into the Central Organization of Salaried Workers (TCO). Appendix 3 shows how fast the white-collar unions grew in this and later periods.

Labor market relations were becoming increasingly complicated as new groups of employees grew in numerical strength and new labor market actors entered the negotiating arena. This caused problems for the LO. There was even competition between the LO and DACO/TCO over three groups of members: civil servants; municipal employees, and restaurant, hotel, and trade employees, many of whom were women. The problem was that there were no clear categories for distinguishing white- from blue-collar employees.

A third important group of collective action organizations from this period is the Swedish farmers' movement. The SAL underwent two crucial reorganizations, changed its name to the National Federation of Agricultural

Cooperatives or SL, and became the peak organization for farmers' cooperatives. A farmers' union organization, the RLF, was created to organize individual farmers, and its founders were inspired by trade unionism. It successfully recruited farmers and became an influential social movement supported by an enthusiastic group of small farmers. Its profile was more adamant or militant than the SL, and initially it did not have a good rapport with government.

Relations between the cooperative and union branches of the farmers' movement were not always harmonious, but they stabilized toward the end of this period. Government pressure was an important reason. Both associations sought government patronage, and financial support was given to the cooperative branch for the first time in 1930. Two cooperative associations, the Swedish Dairies' Association and Federation of Swedish Cooperative Banks, were established with the help of this money. The farmers' movement played an important role in distributing food in World War II. Gradually it became integrated with government and completely dependent on national agricultural policy.

5 The Swedish model

A common element in all definitions of the Swedish model is the ability of opposing political actors to meet and discuss their differences. Today this is called deliberative democracy. Dialogue and mutual respect rather than conflict and mistrust has been the method for interaction among social classes in Sweden. Foreigners fascinated by the Swedish model made study trips to Sweden to discover its secrets. Representatives of the government, LO, and SAF traveled abroad together as evidence of cross-class compromise. Scholars, critics, and the involved actors themselves have given accounts of the camaraderie between labor and capital. All efforts to market labor market peace helped transform the Swedish method for reaching agreement into a mythical model worthy of envy and imitation (Ruth, 1984). Cross-class collaboration and labor peace became part of the Swedish collective identity. One consequence was that Swedes tended to forget that unions and employer associations represented conflictual interests.

Dialogue between opposing political actors became a Swedish institutional tradition in the period of the strong society and was further developed in the present one. It became general Swedish practice for government and collective action organization to meet to solve political problems in a pragmatic, quiet way. Another name for this dialogue between the state and interest organizations is corporatism. It is institutionalization and legitimization of the role of collective action organizations in the political process. This is the Swedish model of politics. This period is dated from the 1940s to the 1950s.

The key prerequisite for favorable treatment of collective action organizations by government was their ability to argue rationally, present expert opinions, and evaluate different policy proposals in a publicly responsible way. A literal translation of the Swedish term for this capacity is 'responsibility for society'. Collective action organizations reacted strongly to any suggestion that they lacked a sense of public responsibility until the sixth period (1980s-1990s), when its value underwent reevaluation. Such suggestions were seen as questioning their legitimacy as public actors and ability to represent their members effectively. The term 'public responsibility-taking' evolved in this and the next period (1960s-1970s) into a rhetorical tool for creating unanimity and stifling dissent within the collective action organizations themselves.

Cold war in the labor market

Normally the term 'cold war' is reserved for the particular warlike but peaceful relationship between the U.S. and U.S.S.R. in the postwar years. Its main characteristics are persistent disagreement between two antagonists, strong bipolarity of the alliance structure, domination of the political arena by major powers, and detente or the development of conflict 'to a more ambiguous stage in which the balance of conflicting and parallel interests is less clearly defined, and the conflict itself has a more diffuse character' (quoted in Gamson and Modigliani, 1971, p. 4). Detente includes cultural exchanges and regulatory agreements.

A cold war involves interaction between two parties. Each power has its own goals, strategies, traditions of action, and expectations, but it is influenced in important ways by the goals, strategies, traditions of action, and expectations of its antagonist. The two antagonists are part of an interminable relationship, which can develop in different ways. There are three theoretical possibilities. Consolidation occurs when a counterpart holds on to what it has and makes its influence more secure in areas where it is already established. Expansion involves extending influence into new areas or increasing it in areas where its present influence is modest. Destruction is attempts to reduce the power of the adversary by weakening it in its strong areas. Destruction is a risky strategy. Unlike consolidation or expansion, it can never be achieved by bargaining and compromise. Two powers do not exist alone in a cold war political environment. Other actors are dependent on the stability of the relationship, and they can be rather successful in persuading the combatants to accept some regulation. Integrative actors and mechanisms handle, mediate, and resolve conflict and implement common goals. Disintegrative forces are intrinsic disagreements between two powers, whose magnitude may change over time.

The labor market relationship between the SAF and LO that began in the early 1900s and ended in the 1970s fits this general description of a cold war. The two powers are the LO and SAF. Integrative actors include the government, political parties, and Labor Court. Integrative mechanisms are labor market rules and regulations, the most important of which was the 'treaty' of Saltsjöbaden. Disintegrative forces are the bipolar interests represented by capital and labor.

Cold war in the Swedish labor market can be divided into five phases (cf. Anders Johansson, 1989a). The LO and SAF developed a common perspective on industrial technology and rationalization in the preparatory phase (1910-31), and cooperation rather than conflict characterized their relationship. The formative phase (1936-44) involved agreement on procedures for discussion and decisionmaking. Government was a particularly important integrative bond in this phase, e.g., in the precursory stage of the Agreement of Saltsjöbaden. The LO and SAF defined their own spheres of interest in these phases. In the enforcement phase (1945-48), Saltsjöbad-like agreements were signed and implemented by the LO and SAF, some of which were applicable into the 1960s. The LO and SAF consolidated their power by centralizing collective bargaining authority to themselves and expanded their spheres of power to include such issues as industrial democracy and the work environment. Contacts between these two labor market giants were frequent in this phase, and neither the SAF nor the LO attempted to use the strategy of destruction. The problematic phase began in the 1960s, when labor market peace was threatened by strikes and politicization, and ended in the late 1970s, when the cold war ended. Destruction became a strategy in this fifth phase.

It is the stories from the enforcement phase that have formed the collective memory of labor market relations. They have been told and retold in and outside Sweden. The mystique of Saltsjöbaden is clearly evident here. At times interests converged between the LO and SAF. Both powers focused on economic growth as well as industrial production and rationalization. The LO deemphasized redistributive economic issues, and the SAF gave priority to developing its labor market expertise and toned down its ideological preferences. Both actors exercised moderation and treated each other with respect.

What has tended to be historically forgotten is that the intrinsic differences between the LO and SAF were at times very apparent. The political debate on economic planning and later the pension issue showed that the interests of labor and capital could be as different as night and day. The two big labor market powers did not always act as monolithic blocs. Disagreements concerned tactics as well as goals. Organized business was not, for instance, completely united internally on what strategy to pursue on the issue of economic planning. Some supported the organized campaign against economic planning (PHM), but the SAF did not want to be too closely identified with it because it did not want to jeopardize its relation with the LO.

Power sharing and public policy

Social democratic tenure did not signify that government only listened to the LO and other labor movement network organizations. A general reason for the success of Swedish social democracy for so many years was its ability to share policymaking power with other political actors. The governing party chose an inclusionary rather than exclusionary strategy that involved cross-class cooperation. It is not very surprising that this period has been called participatory capitalism (see Torstendal, 1991).

This kind of power sharing encouraged the development of a particular political style. Norms for political actors evolved which took the form of consensus and thorough discussion of events and proposals, i.e., 'reaching agreement and reflecting beforehand' (Ruin, 1983, my translation). This decisionmaking style impressed foreign visitors who contrasted 'the absence of crises that produce hastily-considered decisions and the lack of many of the crude forms of obstruction– withholding information, clock-stopping to sidestep legal deadlines, or filibusters' in Sweden with experience from their own countries (Anton, 1969, p. 88).

The basis of Swedish cross-class cooperation is a political opportunity structure that gives collective action organizations a share of political power. The structure is selectively open because certain organizations– particularly labor market ones– are given a central role while others are held at a distance. Selectivity is explained by the class politics character of the Swedish polity.

Defining problems as political

Open political opportunity structures give access to all phases of policymaking. Organizations are, for instance, able to place issues on the political agenda through informal discussions or in written statements and visits to government ministries. Along with public agencies and parliament, collective action organizations help define political problems and initiate policy. All types of organizations use the same general methods of influence, but the selectively open character of Swedish government is underscored by the importance corporatist organizations have given to confidential contacts with government. This development also explains why lobbying never took hold as a political tool for these important organizations.

Power sharing did not imply equal influence. Leaders of organized business believed that they successfully influenced policymaking specifics and details but generally failed to impact its general direction. In contrast, the LO exercised influence on broad, programmatic, ideological welfare policy issues, but its reactions to policy details were basically ignored (see Elvander, 1972). The imbalance here reflects the development of a Swedish social democratic state

(Rothstein, 1986) and social democratic political hegemony. Political hegemony meant that the social democrats defined what issues were political and how they were debated publicly (cf. Pontusson, 1988).

Representation on government commissions

The characterization of selective openness applies to many government bodies. An important example are government commissions, an institution of old origin, that were increasingly used to study how government action could solve welfare problems. Government's willingness to view problems in this fashion shows how collective action organizations convinced politicians that the boundaries of the political sphere should be broadened to accommodate citizen needs.

The composition of commissions ensured that policy proposals were well-anchored in Swedish civil society. Commission members were chosen from political parties and collective action organizations. Organizations was called in as experts or members. The appointment process was selective: certain labor market organizations were the most frequent participants of all collective action organizations. At times collective action organizations directly requested representation on commissions, but government did not always grant them a seat.

Commissions developed into power sharing corporatist bodies. Participation by collective action organizations in them increased after 1925 and skyrocketed between 1935 and 1955. This development led a political scientist to conclude: 'Organizations have replaced the role earlier played by political parties as representatives of the interests of different groups of people within large sectors of societal life' (Meijer, 1956, pp. 83-84, my translation). His comment is interesting for it reveals an important implication of Sweden's political opportunity structure: stable government marginalizes the role played by political parties and parliament and increases the role of encompassing interest organizations and countervailing powers.

Appointment of commissions as well as representation of collective action organizations on them decreased toward the end of this period. By the mid-1960s, commission members from large labor market organizations numbered only two-thirds of what they had in the 1940s and 1950s, and civil servants began to play a more dominant role (Meijer, 1969). This development is an early sign of the fair weather character of Swedish corporatism in that welfare state development, government expansion and intervention, corporatism, and use of commissions have gone hand-in-hand. The system of commissions is an important pillar of Swedish political culture, corporatism, and strong society. It symbolizes the Swedish model.

Official comments by collective action organizations

Another pillar has been the system of official, written comments that gives a larger number of political actors access to government by allowing them to voice opinion on commission and government proposals. Government ministries request public agencies, regional and local government bodies (which may in turn solicit comments from collective action organizations within their jurisdiction before responding themselves), and all kinds of collective action organizations to submit comments. Political parties are excluded from this process as they express their opinions in parliamentary motions. Government began to rely more heavily on collective action organizations in this sounding out process after World War II. Official written comments by collective action organizations play other central roles: they influence long-term government planning and policymaking; political parties refer to them in parliamentary debates; ministers explicitly or implicitly use the viewpoints expressed in them as arguments in government bills, and they help to identify the ideological profile of the collective action organizations for their members, the mass media, general public, government, and other organizations of civil society.

A collective action organization that has not received a proposal for comment has the right to submit its viewpoints, which makes this channel for influence less selective than the system of commissions. Yet the process generally favors collective action organizations with the most resources. At times, collective action organizations in the same sphere submit joint written comments– e.g., the SAF and IF; SL and RLF, and KF and tenant movement have cooperated informally with the LO and TCO. On occasion collective action organizations and public agencies have unofficially coordinated their comments, as has occurred in the case of organized business and the National Board of Trade and chambers of commerce (see Elvander, 1972, Micheletti 1984a).

Involvement of collective action organizations in implementing policy

The Swedish selectively open political opportunity structure has even included collective action organizations in policy implementation. This development is explained by the general character of many Swedish laws. Laws regulating welfare policy are frequently vaguely formulated– i.e., frame laws– and require that the responsible public agency interprets them. Their advantage is that they allow for flexibility in terms of policy objectives and expenditures, which has given collective action organizations a crucial role in policy implementation.

A foremost example of power sharing in policy implementation is membership of collective action organizations on lay boards governing public agencies. The crisis situation of World War II laid the groundwork for the creation of lay boards, which grew rapidly in number in this and the next period of civil society history. War crisis and the maturity of collective action organizations are important reasons for this development. Lay boards were created without much political contemplation. They fill a need for policy expertise, special interest representation, and civic assessment of the soundness of policy. Government has included collective action organizations as board members when their acceptance of agency decisions has been viewed as crucial for successful policy implementation.

Collective action organizations represent the second largest group of board members. Members are selected by the government; organizations not selected but requesting seats are usually denied representation. Many organizations have been represented on lay boards whose tasks affect their members directly. Also, it has been common Swedish practice to include countervailing powers or collective action organizations that balance one another. For instance, if the LO is a member, the SAF is usually one as well. Government has even created countervailing powers for this purpose, as exemplified in the case of consumers representatives on agency boards in the agricultural policy sector. Representation of countervailing powers on lay boards and in other government bodies has been a defining trait of Swedish corporatism (Micheletti, 1991a). Decisions of the lay boards have usually been compromises between the different collective action organizations, which have reflected their ability to assume public responsibility and mediate between the interests of their members and government.

Another example of selective power sharing is delegation of public responsibility to collective action organizations. Organizations have been given authority to implement public policy and have been reimbursed economically by government for it. Frequently they have been able to exercise a considerable degree of discretion in their work. For instance, farmers' associations have been delegated responsibility for certain aspects of agricultural regulation, e.g., milk regulation. Delegation of public authority played an important initial role in institutionalizing and legitimizing farmers' associations.

There are several other ways for collective action organizations to influence public policy. Representation on national advisory boards, courts, and regional and local bodies are a few examples. Many organizations have intimate contacts with government ministries, public agencies, and other administrative bodies. Special networks of collective action organizations, government officials, and civil servants have developed within almost every policy sector.

Power sharing between social democratic government and collective action organizations benefitted Sweden in many ways. There was labor peace and welfare state development. Power sharing also had its costs for all involved political actors. Yet there was really no public and academic discussion about the democratic problems that it created until quite late in the period. Rather, emphasis was put on the positive aspects of corporatism for the Swedish nation. Two general problems of power sharing are explored here.

All participants in the networks that developed in public policy have tended to perceive policy problems and solutions in similar ways. A kind of group think evolved among the participants. Ideas suggested by people outside the sectoral network were frequently rejected because they did not fit into the common perception of policy means and aims. Newer participants in the network were socialized in the sector ideology. Over time this led to problems of governability and internal, democratic problems for collective action organizations. For instance, it happened on occasion that organizational representatives had difficulty understanding and accepting policy demands from their members which differed from the sectoral ones.

Another serious problem of power sharing was the minor role played by political parties and parliament. Some background information on the political setting is needed to explain this development. The cause here was closer contact between government and labor market organizations that developed after the 1948 parliamentary election. This election returned the social democrats to office with a minority government, as a coalition with the Agrarian Party was not possible. The social democrats needed the cooperation of organized farmers and business, which took the form of selectively open deliberative sessions. The theme of all meetings was Swedish economic and welfare policy. The government's aim was to ensure governability and improve relations with organized business, which generally lacked confidence in the social democrats. This attitude was viewed as very serious, as government leaders knew that they would be reliant on business associations to make and implement public policy.

The first more permanent deliberative effort became known as the Thursday Club and existed between 1949 and 1955. Participants came from Sweden's power structure elite– organized labor (LO and TCO but not SACO), organized farmers (SL and RLF), and organized business (SAF and IF). Political parties were excluded. The club functioned as a 'complaint desk' for organized business (Elvander, 1972), but once its number of participants grew it ceased to function as a corporatist tool.

Contact between the government and organized business continued in a different setting, at Harpsund, a residence of the prime minister, from the mid-1950s to 1964. Organized business and the LO were initially the only collective action organizations invited to deliberate, but the TCO and KF were later included as were other collective action organizations, which participated in certain sessions. This time exclusion of the political parties of the parliamentary opposition led the leader of the Liberal Party to coin the phrase 'Harpsund democracy', meaning decisionmaking on political issues outside parliament. His outcry started a lively public debate on the democratic problems of power sharing. Many non-socialists believed that social democratic leaders used the Harpsund setting 'to play Swedish business off against the parliamentary parties' (conservative leader Gunnar Heckscher quoted in *Arbetet*, 15 Feb. 1969, my translation).

This is one of the first times in civil society history that corporatism was seriously questioned. The liberal party leader overstated his case; research later showed that the deliberative sessions were more similar to free discussion sessions for the exchange of information and ideas than decisionmaking forums (Elvander, 1972). Yet his concern for the development of Swedish political democracy showed his foresight as a politician. Criticism of corporatist exchanges became an important political issue in the next two periods. The Liberal Party was a foremost critic. Quite possibly the sessions ended because the government found the criticism brought up in the debate on Harpsund democracy to be burdensome and troublesome. Perhaps organized business and the government had a premonition about the rise of the zeitgeist of the 1960s and desired to deliberate with each other in a more secluded setting out of reach of the mass media. Such a setting existed in a council on economic planning created in 1962. Participants included organized business, farmers, and wage earners as well as other collective action organizations involved in welfare state reform.

Partisan politics and power sharing

Power sharing within the framework of the Swedish model and cold war in the labor market did not mean that opposing actors agreed politically and stopped struggling for power. There were major and minor skirmishes. The largest labor market conflict since the general strike of 1909 broke out in the engineering industry in 1945. The other important conflict was ideological and involved an earnings-related (supplementary) pension plan (ATP). The issue divided Sweden into a socialist and non-socialist camp in the late 1950s and caused internal cleavage within those collective action organizations with member sympathy divided among the political parties. A case in point is the TCO, whose public- and private-sector member organizations held dissimilar views on the issue (Micheletti, 1985b).

Disagreement on pension plans terminated the social democratic and agrarian coalition government, which ruled Sweden from 1951 to 1957, and a national referendum on the issue in 1957. The public chose between three referendum alternatives which ranged across the entire spectrum of Swedish political ideology. The issue itself can be seen as a part of the larger ideological debate on the planned economy that began in the period of the strong society. Organized business did not take a unified stand in the debate on the planned economy, but it was not hesitant about challenging the LO on the pension issue. The social democratic alternative calling for a state-run obligatory pension system received a plurality of votes and was later enacted in a close vote in parliament.

Power sharing had interesting implications for Swedish politics and particularly for the behavior of collective action organizations and political parties that were not part of social democratic political hegemony. The Liberal and Conservatives Parties were generally inclined to take strong ideological positions and challenge the system of corporatism, while organized business chose caution and opted for the power sharing route to exercise political influence. The Agrarian and later Center Party positioned itself in the middle. The positions taken by the parties are explained by the roles played by the political actors in the corporatist exchange. Organized business was included; the liberals and conservatives excluded, and the Agrarian/Center Party shared power with the social democrats and guarded the corporatist status of organized farmers.

Expanded government and marginalization of collective action organizations

Social democrats were the main driving force in changing the boundary of the political to include social citizenship. The result was the 'expanding state' (McEachern, 1990) and one of the largest public sectors in the world. Government expansion meant intervention into new spheres, promotion of institutionalized negotiations or power sharing between class organizations and the state, and creation of 'public administrative corporatism' (Rothstein, 1992). These developments are captured in the terms the 'strong society' and 'Swedish model'. Government commitment to social and economic citizenship impacted collective action organizations in many ways. Probably the most important one was the transfer of social services from them to local government, a development which has been called 'municipalization', a kind of nationalization of welfare service production to local government bodies. Few scholars have analyzed the effects of municipalization on social movements, but it is clear that transfer of services often led to their marginalization as socially useful volunteering associations.

Almost all collective action organizations provided welfare services. A network association to coordinate their efforts even existed. Collective action organizations were responsible for career guidance and job placement for women, shelters, nursery schools and training of pre-school teachers, public libraries, sick insurance schemes, and home help (Svensson, 1962, Eriksson, 1962).

Once government decided to take on a social problem it usually chose to municipalize or nationalize services. Municipalization of these services led to a dramatic increase in the number of public sector employees, many of whom were women, and the rapid growth of public sector unions. However, there are few insightful exceptions. They are the youth and housing problems.

Care for young people highlights clearly the political thinking of this period and general Swedish attitudes about the virtues of membership in collective action organizations. Debates were held on whether government or collective action organizations should take responsibility for juvenile delinquents. Proponents of local government involvement argued that youth associations belonging to established collective action organizations could not reach problem teenagers, as this group was generally not members of collective action organizations. Opponents of municipalization, many of whom were organizational leaders, believed that, given the proper resources, established collective action organizations could do just that. They pointed to studies showing that established collective action organizations were one of the foremost remedies for juvenile delinquency. It is important to remember here that established collective action organizations had become dependent on government patronage and did not want to lose this financial resource by municipal take-over of youth services. They received considerable sums of money for their youth associations at an early date; the practice was institutionalized by the end of the 1930s, after considerable pressure from the organizations themselves.

Proponents of municipalization questioned traditional Swedish thinking about the virtues of membership in collective action organizations. A well-known commentator with roots in the labor movement wrote in a book that troublesome teenagers found community centers run by many collective action organizations boring and old-fashion: he likened them to churches and temples with 'awkward church-like meetings and the weighty moralism of popular education' and finished by castigating local politicians, popular educators, and priests who believe that the general public benefitted from 'people [being] shoved into organizational boxes' (Tage Lindbom quoted in Olson, 1992, p. 142, my translation).

Government commissions were appointed to sort out the problem. Parliament decided that municipalization was not a good idea and declared that government involvement should be restricted to supporting the efforts of established collective action organizations. As a consequence, government patronage to youth associations increased as did the number of collective action organization involved with problem children and teenagers. Established collective action organizations began to view young people in a new light. They ceased being a potential organizational resource and became a problem group to be disciplined and reared. Youth associations were given a new role. Previously they had focused on socializing young members in the ideology of the organization, but now they were semi-governmental reformatory institutions:

> Government was not particularly concerned about the classic role of social movements to transform society. It did not fund collective action organizations for their ideologies. Rather, collective action organizations were employed as *agents* to instill the non-associated youth with acceptable public values. This was the interest of the state, and it explains why government patronage was granted (Olson, 1992, p. 194, my translation).

Government expanded its control over collective action organizations in this area through economic steering, and organizations extended themselves in ways which deviated from their traditional services. Few people at this time reflected on the implications of public responsibility-taking and acceptance of government patronage for collective action organizations.

The housing problem was also initially excepted from municipalization. As in most countries, housing shortage was a major social problem, and many different social movements– in particular the labor and emerging tenant movement– were active in mobilizing public opinion on the issue and demanding local government action, which occurred to a limited degree. A homestead movement emerged as well, with the active participation of different movement organizations and private companies. Yet it was the housing cooperative movement and particularly the HSB that did the most in the late 1920s and early 1930s. The HSB worked together with several municipalities, and this model of cooperation was later adopted by other countries. Government became more involved in housing policy after 1945 and municipalization occurred to some extent, but the HSB and other housing cooperatives remained during this period the main providers of service. Membership in HSB increased steadily and more than tripled between 1930 and 1945 (Johansson, 1962, see appendix 6).

Creation of a large, expensive public sector to deal with social welfare issues has, with few exceptions, epitomized the Swedish way of solving welfare problems. Efforts by collective action organizations to make people aware of social welfare problems and dedicate themselves to their eradication put the issues on the political agenda and changed the boundary of the political.

Political agreement on the need to solve a social problem came to be equated with government action in postwar Sweden. Collective action became synonymous with government action, and this led people to conceive of government as a service organization (Hancock, 1972).

An important development here was the integration of collective action organizations and government. Organizations participated on government boards once their welfare services were taken over by government. Public administrative corporatism developed in many sectors. The strong society evolved into the good society, and at times it seemed as though Swedish citizens believed that the welfare state could alleviate all problems of human existence. This viewpoint influenced social science as well. The best example is a scholar of social welfare in the Nordic countries who not only studied material welfare (having) but even the sense of solidarity/community (loving) and individual integrity, empowerment, and efficacy (being) (Allardt, 1977).

Many collective action organizations were willingly or unwillingly marginalized by the welfare state project. The public space that they had carved out was now occupied by the burgeoning welfare state and social partnership between government and labor market organizations. Municipalization impacted collective action organizations severely: members and supporters had less money to contribute to their organizations because taxes had been raised to finance welfare services; organizations were forced to find other sources of financial support, and government patronage became more important, even for the free churches whose independence from government was a foremost organizational goal in earlier periods. Also, municipalization meant that many collective action organizations lost important contact with social reality, and this led to declining member commitment.

Not all social movement organizations viewed the expanding state as a sound, democratic development, as poignantly illustrated in the following excerpts from speeches by two leading free churchmen:

> The modern state has also become a dominant part of our lives. It believes that it owns us: it claims to form our lives and even to decide our religious beliefs. And it comes very close to Christian claims with these aspirations of power....[We] cannot allow the state by itself to work for abstinence, civic spirit, morality, social consciousness, etc. Free churches must show that they are prepared to take responsibility for civic education (Sollerman, 1958, pp. 10, 17, my translation).

> [Some people react by viewing] the development of organization-Sweden as a institution of redemption for contemporary Swedes, i.e., people seek blissfulness through their own achievements, and it has even been said that interest organizations have become analogous to

> the Catholic Church of the Middle Ages, as an asylum for self-redemption of sorts or a kind of machinery for achievement. People have become trapped in interest organizations and need to be liberated from them (Källstad, 1964, pp. 19-20, my translation).

Problems with organizational renewal and internal democracy

The period of the Swedish model marks a clear change in Swedish civil society. The external setting for collective action organizations was dramatically transformed, and changes were occurring within collective action organizations themselves. They had now reached the life cycle stage of full maturity; some were even showing signs of stagnation. Members began in very modest, restrained ways to confront their leaders and question the democratic traditions of their organizations. This criticism would grow stronger in later periods. Also, the organizations which had experienced problems with organizational maintenance in earlier periods found that their situation was as– if not more– severe as Sweden approached the 1960s.

Swedes sought rather to solve social welfare problems through social engineering and large-scale structures, planning, and construction. This was the epoch of grand programs, Sweden's great society (Wittrock and Lindström, 1983). Foreign observers saw Sweden as a progress machine and 'the prototype of modern society' (Tomason, 1970). Most collective action organizations accepted and adapted themselves to this mechanical, material interpretation of social progress. Organizations as diverse as unions, free churches, adult education, and teetotaling societies as well as the National Tenant Association merged their structures and standardized their operations. Bureaucratization, centralization, and universalization of services typified the approach adopted by actors involved in the Swedish model.

Yet while many of these projects successfully solved the material problems of social welfare, they discouraged individual diversity and nonconformity. Efficient bureaucratic solutions created a situation of 'overconformity', where rules emerged as goals of public policy, and social democratic spokeswomen Alva Myrdal issued an early warning on the dangers of bureaucratic development when she coined the term 'St. Bureaucratius' (see Ahrne, 1987, pp. 191, 197). The Swedish word *lagom*, meaning 'just right' for everyone, and the 'law of jante', proclaiming that solidarity and unity forbid individual excellence and thereby diversity, have been advanced as linguistic indicators of the political culture of the Swedish model (Sandemose, 1984). This conception reached Swedish popular culture. The lines of a popular song in the heyday of the Swedish model were 'You should do as the Svensson's do and not differ from the collectivity'. Solidarity with the cause of collective action organization was replaced by loyalty to the organization itself.

Dissatisfaction with the view of individuality that was held by the LO, one of the main proponents of the Swedish model, was expressed in novels that appeared at the end of this period. One author lamented the inflexibility of collective action organizations and the oligarchical tendencies within unions. The novels of another author, which strongly criticized tendencies toward centralization, authoritarianism, and oligarchy within the LO– as illustrated in the description of the LO as a royal court complete with princes, who persecuted a young man for criticizing collective affiliation and questioning union doctrines– as well as the union's preoccupation with material values (Salomonson, 1959 [1991, p. 553]), met with deep disapproval among social democrats and trade unionists. Arne Geijer, President of the LO, spoke rarely and only disparagingly of the author and his books, and it has been reported that he banned the books from the LO. He maintained that all tendencies toward oligarchization within the LO were counteracted by a vital union democracy at the local level.

Yet the first systematic opinion survey of union democracy showed the embryo of problems within the LO. Different categories of members from one growing LO union, the Swedish Municipal Workers' Union (SKAF), were questioned in 1958 (Lindblad, 1960). An overwhelming majority had a positive impression of their union, but it is of interest that about 20 percent were skeptical and that a small, statistically insignificant number of members in larger cities were more negative than members from other areas. This development would become more pronounced in the final two periods. Younger members were even found to be more negative to unionism. The political scientist who authored the study had difficulty explaining this development; but novelists of the time understood it as a consequence of the political culture of the Swedish model.

The study singled out another problem area: members lacked commitment to the union. About 40 percent did not participate in union affairs; women were three times more inactive than men, and about two thirds of members under the age of thirty were disengaged. No appointed official was female, and only a sixth of local and central elected officials were women. Yet about 20 percent of the membership was female. No attention was paid to the role of immigrants, a group of members which began to grow in number and experience union-related problems. Also, union mergers had not solved the problem of democracy and structural size: participation had decreased greatly and dissatisfaction increased dramatically in the largest local union units, when compared to small units with less than one hundred members. The study even showed that leaders generally did not know where members stood on various union issues. The LO President tried to rationalize the results, and the SKAF did not have the foresight to deal with them. The problems reflected in the survey would become very serious in the next two periods.

Other collective action organizations had similar experiences. Members were much less committed to movement causes. This was an early warning sign about a shift in involvement from collective to more private action (see Micheletti, 1994). Organizations were acutely aware that they needed to renew themselves by recruiting newer and younger people. Their members were growing old. Drops in membership rates within many collective action organizations can to a large degree be explained by inability to replace deceased members with new, living ones. Young people became an even more attractive target group. Collective action organizations pondered how new groups of members could be recruited. Various attempts were made to recruit women and activate them as members.

Women had been more active participants in collective action organizations in World War II, but mergers of local associative units in the postwar years marginalized them. They found it more difficult to assert themselves in new, larger settings. One woman expressed her frustration succinctly: 'Men are such egoists; they think they are the only people capable of assuming organizational duties' (quoted from Stolpe, 1945, p. 10, my translation). Special women's units were set up as were women's network structures. Women were considered important because they filled the ranks of members and performed housekeeping tasks for the organizations.

The free church movement had the opposite problem: it needed men. There were two female members to every male member by this period. This led to private problems for members, as the rather isolated life style of free church sympathizers meant that marriage with non-believers was either impossible or forced the believer to quit the movement. A study of free church congregations in Uppsala showed that people in the age category of sixty years or older had increased from 8 percent to 28 percent between 1940 and 1960 and younger people (thirty-nine years or younger) had decreased by 10 percent. Most older members were women (see Nyhlén 1964, Palm, 1963). The issue was literally the ability of the movement to reproduce itself. This trend was broken somewhat in later years.

Many organizations admitted that their traditions and methods of action were old-fashioned and unacceptable to new groups of members. The use of PR and role of the mass media were frequent topics for internal debates and reports on organizational strategy for the future. Associative membership and more intense collaboration with other collective action organizations were suggested again and frequently tried, but it was also feared that such solutions would recast organizational goals and alienate traditional member groups. Some leaders even questioned whether organizational goals and ideologies still appealed to the broad public. Temperance vs. teetotaling as the proper organizational goal was debated within Verdandi and the IOGT, with the issue being whether more people could be recruited if moderate drinkers were allowed as members. The Verdandi decided to drop absolutism from its constitution, which meant that it no longer was classified as a teetotaling organization. The long-term effect was an increase in members. That certain collective action organizations were successful in recruiting and keeping members because they offered selective incentives led other organizations to wonder whether member service was not the wave of the future. For instance, the teetotaling insurance organizations continued to attract members long after other movement organizations began to decline (see appendix 2).

This was the predicament facing Sweden's collective action organizations. Idealism, voluntarism, and devotion of free time to organizational tasks no longer attracted large groups of citizens. Collective action organizations had in many ways both won and lost the struggle for their causes. They had changed the boundary of the political but could no longer instill their sympathizers with a sense of urgency for their issues and causes. Members found nothing really important to protest about and were content with the development of politics and society. Swedes had become almost completely reliant on government for collective action. The public was simply bored with or alienated by the ideology, message, and activities of many collective action organizations. Organizations had to compete with the modern entertainment industry and other, newer collective action organizations for public attention. Soon television and later mass media in general would threaten to replace the role played by older movement organizations.

6 Rethinking politics and the dissatisfied generation

The Swedish mentality and political landscape changed greatly from the 1960s. Members, students, intellectuals as well as politicians and government officials fervently criticized the presence, political function, and goals of established collective action organizations. New left and post-material values found a stronghold among young people and in other, newer groups which seriously began to question the basis of the Swedish model. This was the period of anti-American sentiment, solidarity with third world countries, new socialist ideology, and most importantly the entrance of postmaterial political values and emergence of new social movements. This period is dated to the 1960s and 1970s.

There were extremely high levels of collective action by grassroot groups, public demonstrations, confrontation with the police, political deviance, and general despondency with the established political order. The term 'dissatisfied society' characterizes this collective sentiment of disappointment with political, social, and economic development (Heller and Fehér, 1988). The collective action organizations that channeled dissatisfaction in Sweden were similar to those in other western countries: the student; anti-war protest, and counter-culture movements of the 1960s and anti-nuclear power movement of the 1970s. One central target of criticism was social democracy. It was viewed as a monolith, a hegemonical power holding its grip on political power and Swedish society. Capitalist ideology in general and Swedish capitalism in particular were other targets.

Social democracy and social movements

Swedish social movements and social democracy have had the same, basic historic task. They wanted to change the boundary of the political by including more aspects of life in government responsibility. Many movement protest actions were initiatives for government action. The result was the transfer of tasks from movements to government and a loss of autonomy for collective action organizations *vis á vis* government. Few social movements challenged this development.

Political actors were pleased or frustrated by the social democratic character of expanding government and the party's hold on the nation's political identity. People reasoned differently about the desirability of a united, strong labor movement with several collective action organizations as allies. Non-socialist collective action organizations were becoming frustrated with the might of Swedish social democracy. In the late 1960s, the SAF began to reconsider its strategy of cooperation, expertise, and non-politicization, which it had adopted in an earlier period, and sought to dissolve the glue of social democratic and union hegemony. This trend-breaking decision was in part a reaction to the emergence of new left groups of the 1960s. The SAF's resolve on this issue became firmer in the 1970s, when the LO and TCO decided to improve industrial democracy through parliamentary legislation rather than labor market negotiations. The SAF called the 1970s 'the decade of devastation' (Schiller, 1988), a label which expressed its disappointment with general political development, failure of the non-socialist governments to erode social democratic hegemony, growing strength and breadth of the unions, and weak political position of organized business (Micheletti, 1984a, De Geer, 1989). The SAF began to evolve from a labor market organization disciplined by expertise and bureaucratic professionalism to a protest movement for organized business employing media tactics and strategies.

There were other reactions by non-socialist sympathizers as well. A foundation called the Forum for Non-Socialist Debate was established in the 1960s and published several reports on the social democratic apparatus of power. The authors, some of whom were academics of non-socialist persuasion, focused their attention on social movements not formally allied with any particular political party but commonly associated with the Social Democratic Party. Among them were the largest organization for senior citizens, the National Association of Pensioners (PRO) and the National Tenant Association.

The studies were set up as social science investigations of the degree of integration between collective action organizations and the Social Democratic Party. Their results were predictable: movement leaders were shown to sympathize with the party and function less as representatives of member interests and more as 'the [social democratic] government's foremost spokespersons' (Norberg, 1966, p. 93, my translation). Research results from less eager and partisan scholars confirmed a social democratic bias in many collective action organizations and the difficulty, inability, or lack of interest of sympathizers of other parties to assume top leadership posts in, for example, the Swedish Union of Municipal Workers (Lindblad, 1960, SKAF, 1973).

The long tenure of the Social Democratic Party and inability of the non-socialist parties to break its hegemony even led this foundation to commission studies of social movements collectively affiliated with the party. Their frustration with social democratic hegemony was clearly expressed in the preface to a report:

> With few exceptions only social democratic sympathizers hold leadership and administrative positions locally and nationally in social democratically governed organizations as the ABF, HSB, Cooperative Housing Union, *Folksam* Insurance Company and *Reso* [an automobile insurance company], and Labor Movement Community Center. Frequently they hold office because they are competent and capable; in other cases their foremost merit is membership in the Social Democratic Party. It is not really possible to know whether these functionaries devote all their time to politics, how much time they devote to it, or whether they do so on work time. But it cannot be refuted that social democracy has access to a large cadre of politically-conscious and socialist-oriented functionaries outside the party organization. These functionaries and organizations under their control comprise the political part of the social democratic labor movement, i.e., the social democratic power apparatus. This combination of economic and political power is mutually supportive. As national and local office-holders, the political leaders give economic grants– i.e., taxpayers' money– to social democratic companies. Economic strength is marshalled in this way, and it can be– and is in fact– used for ideological purposes, e.g., to finance the political part of the labor movement.
>
> No other such consolidation of economic and political power exists in Sweden. The role of the apparatus is to maintain its own strength, which is based on social democratic political majorities....Any change in political power arrangements is greatly inhibited by intense pressure from the social democratic power trust. Yet such change is important for the further development of democracy (Anners, 1966, pp. 5-6).

Accusation of alliance with the Social Democratic Party was an extremely sensitive issue for certain collective action organizations. The nonpartisan TCO had the most trouble in this regard. Editorials in non-socialist newspapers, argued incessantly that the TCO, due to its stand on wage earner funds (an extremely hot political issue in the late 1970s), was a fellow-traveler of social democracy. This caused the TCO worry for over half its members voted for other parties and many opposed wage earner funds. Time-and-again the TCO tried to convince members and the public that it was not a social democratic organization. Yet on certain important occasions its actions belied its words (Micheletti, 1985b).

Non-socialist spokespersons began to see corporatism and the presence of politically influential collective action organizations as evidence of social democratic hegemony. Rather ironically, authors of books and reports from the developing opinion formation apparatus of organized business– a special interest in itself– argued that special interests and social democracy were mutually supportive. Critics of social democracy believed that collective action organizations were general impediments to economic, social, and political progress. Organized business, the Liberal and Conservative Parties, as well as the Center Party to some degree, had become obsessed with the issue of social democratic hegemony and power of unions and other collective action organizations which were part of it.

Far left and new left groups were also displeased with the political landscape of Sweden. Both non-socialist and leftist groups believed in what may be called a conspiracy theory of social democratic hegemony that involved assumptions about the provision of government money to programs aimed at promoting social democratic power over Swedish thinking. The quotation from the Forum for Non-Socialist Debate shows this clearly as does an article in a leftist journal, which concluded that the ABF received government grants for language training for immigrants to brainwash immigrants into accepting social democratic rather than communist ideology (Bodegård and Gustavsson, 1974). Both political groups saw social democracy as an Orwellian big brother, whose persuasive power was so strong that most Swedes were entirely unaware of its existence (cf. Huntford, 1971). The groups worked independently of each other but the sense of being stuck in the glue of social democratic hegemony is the common root of their political frustration.

Social democratic collective action organizations took the criticism calmly and did not feel threatened by it. The Swedish Union of Municipal Workers proudly stated that the labor movement was the best equipped of all collective action organizations and that it covered most sectors of society in its organizational strength. Yet this confidence belied a loss of voting strength for the party, with the 1966 municipal election as one of its worst in modern history, and an ongoing contest over government office, which culminated when the Liberal, Center, and Conservatives Parties formed a coalition government in 1976.

New politics confronts old politics

Traditional or old politics in Sweden not only concerned the use of standard solutions to satisfy material needs and wants but also involved a special conception of democracy based on representative principles, procedural concerns, and a hierarchical, pyramidal power structure. This shared conception of democracy created a strong, general bond between government and established collective action organizations. It 'cemented' different parts of Swedish society together (see Elster, 1989) and provided a "'tool kit' of habits, skills, and styles" from which Swedes constructed their reality (Swidler, 1986, p. 273).

Elections are central to this conception. Officials reflect member interest *because* they hold elective office. All structural levels are included in this process. High-level elected officials are seen as being able to reflect correctly the needs and desires of grassroot members because they are elected by them via representatives on other levels.

A second key element is the decisionmaking technique of anchoring of decision. Anchoring of decision remains an elusive concept. It has been a decisive argument in defense of the representative and democratic nature of organizational proposals, decisions, and actions, but the concept itself does not appear in their constitutions. Organizational officials cannot even really explain the process of anchoring, but mere mention of it has thwarted criticism. It has given representatives leeway to formulate policy stands on the basis of how they choose to synthesize preferences, values, and beliefs expressed only indirectly by members. Anchoring of decision has signified that leaders can avoid direct accountability to their members for policy stands and actions. It has been a central element of Swedish elite democratic practices and characteristic of corporatist organizations (Micheletti, 1985b). Elite democracy and corporatism as well as the Swedish model and old politics have gone hand-in-hand in Sweden.

Another crucial aspect of this conception of democracy concerns arenas and procedures for debate, discussion, and disagreement. Almost all collective action organizations hold congresses that set ideological priorities, elect top leadership, and debate organizational issues. The opinions of lower levels of the organizations are expressed in the form of motions, which either introduce a topic for debate or respond to a policy proposal formulated by appointed and elected officials. Standard operating procedures and representation at congresses have central importance. They distinguish for Swedes what is from what is not democratic practice. Collective action organizations must follow this formula if they want to receive government patronage. This democratic norm even convinced the SAF in the late 1970s that it should hold congresses, which are only advisory but fill a useful public relations function, and call itself a social movement (Micheletti, 1988b).

Democracy based on representation of groups and decisionmaking at periodic congresses served Sweden well in earlier years, but its usefulness declined in this turbulent period of civil society history. There were two general weaknesses: (1) democratic representation based on allocation of votes to structural groups undervalued the actual impact of diversity of opinion for the organizations themselves and importance of open discussion for democratic development and organizational renewal and (2) the primacy of congresses as final decisionmaking bodies led to policy inflexibility and inability to adapt to sudden changes of central importance. Examples of heterogeneity are the growing number of women and immigrants in the workforce, which implied a need for more openness to their ideas, and the declining importance of industrial production for employment, which meant the transfer of increasing numbers of wage earners to post-industrial production, i.e., the public and service sectors. Probably the best examples of the actual consequences of the breakdown of representative democracy are the wave of wildcat strikes and struggles over nuclear power and wage earner funds.

The sentiments of dissatisfied workers began to heat up the cold war in the labor market when a series of wildcat strikes threatened the Swedish model in 1969 and 1970. The strikes were not sanctioned by the union, took place while collective contracts were still in effect, and therefore were illegal. They were not the first wildcat strikes in Swedish postwar history or the 1960s, but they were larger and more serious than previous ones (see Korpi, 1974). It is worthy to note that miners, a group of people singled out in the literature as important in most protest waves, conducted a strike at the modern, government-owned LKAB mines in Northern Sweden in December 1969. Their action shook the nation and received national and international mass media coverage. A total of 5,000 workers participated in the strike, which made it the largest strike since the metal workers' conflict of 1945. Miners expressed new political values when they voiced their dissatisfaction with working conditions in such slogans as 'We are not machines' and 'Give us back our human dignity' (Hirdman, 1988, p. 308, my translations).

The strike was not only aimed at the mining company or caused by dissatisfaction with wage levels and working conditions. Strikers were even dissatisfied with their central union, the Union of Mining Workers. They saw it as playing the role of patrolman for social democracy. They viewed the central union as elite democratic and criticized it for signing wage contracts without consulting the locals. Members were also increasingly alienated from union locals, which they saw as controlled centrally and whose meetings were poorly attended. The first reaction of the LO and Union of Mining Workers to the strike was defensive, and they put the blame on communist unionists. But the real cause was confrontation between old and new political values. The strike committee adopted participatory democratic principles and involved all strikers in important policy decisions. Many strikers believed that established union practices could not help them reach their goals. Union officials felt threatened by this new order, as it went against traditional union democracy based on representative principles. It ended after almost two months. Wage policy was changed; a new, more open system for collective bargaining was introduced, and union locals were given more authority. The activism and enthusiasm of the strikers diminished rapidly afterwards. Though it was decided that the strike committee would continue as a countervailing power, it dissolved shortly afterwards. The sense of urgency had disappeared.

The LKAB strike was extremely important for Swedish politics and labor market relations. It encouraged protest at other worksites and led unions to reconsider democratic centralism and stress industrial democracy, a topic first put on the trade union agenda in the 1950s. It even played a role in the 1970 parliamentary election. But most importantly, it showed Swedes that it was possible to question traditional structures and practices and put new demands on established collective action organizations. Authors of a book on the LKAB strike had foresight when they concluded:

> The hegemony of the social democrats is a major characteristic of the Swedish union movement. This united demonstration of strength by the Social Democratic Party and LO has in various situations been a force that has made it possible to assert the interest of workers, e.g., the satisfaction of worker demands for pension rights through legislation and state action. What we need to question here is how this force has impacted the ability of workers to assert their interests at the grasroot level (Dahlström et al., 1971, p. 204, my translation).

Sweden lost its international reputation for labor peace in this period. Confrontation between new and old politics is an important general explanation here.

Labor war broke out in 1980 shortly after the public referendum on nuclear power when almost 720,000 wage earners were directly involved in an eleven-day strike and lockout that paralyzed Sweden: 689,000 privately and 28,000 publically employed wage earners were involved; 699,000 employees were locked out. The strike reminded Sweden as a nation how dependent she was on good labor market relations. Decisions to take just a few key union members out on strike were immobilizing: the Stockholm subway stopped running; citizens panicked and stockpiled household necessities. The conflict has been likened to the general strike of 1909 (Schiller, 1981).

Immediate reasons for the conflict were: economic imbalance and uncertainty over the government's general economic policy; problems with coordinating the 1980 collective bargaining sessions; calls for union and worker wage restraint and public responsibility-taking; a long, tedious collective bargaining session that began the year before, and a number of wildcat strikes that gave expression to general worker dissatisfaction over wage levels. Less immediate reasons are dissatisfaction among many trade unionists with the non-socialist government and the general desire of the SAF to challenge union power and question public sector growth (Broström et al., 1981). By this period, the SAF had become a public proponent of certain neo-liberal political values and wanted to change the boundary of the political in Sweden by delimiting government.

Also, there were important changes that made traditional Swedish collective bargaining more difficult. The public sector had grown to be one of the largest in the world. By 1980 it employed 34 percent of all wage earners and 44 percent of all working women and accounted for a growing proportion of Sweden's gross national product (SCB, 1990). White-collar unions had evolved from a 'fellowship group' into a true union movement with closer affinity to the LO than employer groups (Nilsson, 1985). Together with a legislative act from 1966 that gave public sector unions the same basic right to strike as private sector employees, their numerical strength and new policy resolve on union demands meant that the Swedish model of collective bargaining had changed forever (Elvander, 1988). More labor market unrest could easily be foreseen; the bipolar cold war in the labor market had become multipolar and much less cold, and the foundation of the Swedish model began to rock.

New politics confronted old politics in other ways, for instance over the nuclear power issue. Postmaterial thinking vied with conventional conceptions of material welfare in this struggle. Disunity even forced a non-socialist government to resign (Petersson, 1979, Vedung, 1979). Soon it became very clear where collective action organizations stood on nuclear power, but many collective action organizations found that their members were divided on the issue. Not even all environmental organizations were united: traditional concern for conservation did not always coincide with environmentalism. Union members opposed to nuclear power found themselves in conflict with leaders who criticized them for holding irresponsible views on economic and employment policy, challenging standard procedures for union democracy, and doubting how well leaders represented their members on the issue. The TCO had the most problem in this regard, and the somewhat successful protest group TCO Members Against Nuclear Power showed clearly that traditional union democracy had difficulty in dealing with watershed events that suddenly changed the parameters for debate and gave sustenance to new political issues (Micheletti, 1985b).

New politics reached Sweden in the 1960s but its impact was more pronounced in the 1970s. The American author who coined the term 'the middle way' in the 1930s to express his admiration with the successful Swedish combination of collectivism and individualism returned to Sweden in the 1970s to review the middle way. His verdict was that the middle way was on trial (Child, 1980).

Emergence of new social movements

The history of civil society in Sweden shows how needs of citizens have led to the establishment of collective action organizations. Dissatisfaction with old politics led people to mobilize more resources for old collective action organizations that captured the sentiments of the decade. New branches of old social movements as well as ad hoc campaign and new collective action organizations emerged. New social movements comprise the fourth wave of organization. Their issues are war, gender inequality, ethnic discrimination, ageism, ethnicity, environmental pollution, elite democracy, central government, and neglect of global solidarity. Their struggle was human integrity, and their enemy was destruction of humanity. Examples are the new women's, international and humanitarian, environmental, anti-nuclear power, new peace, NIMBY (Not In My BackYard), and new student movements. Young people and women became a political factor of importance. More than any other group, they confronted old politics with new political thinking. They encouraged Swedes to rethink progress by democratizing differences, i.e., to accept diversity and new political actors. Their cause was very similar to that of collective action organizations of the first wave. Both waves represent organizations that wanted to sign a new social contract and rethink Swedish democracy. Another similarity is the ability of both organizational waves to attract the youth.

Rather surprisingly, dissatisfaction among union members did not lead to attempts to establish new union organizations. The only exception is the creation of the Swedish Dockworkers' Union in 1972 as a reaction to policies within the LO and the expulsion of dockworkers from the Union of Transport Workers. The unions' hold on politics and members had been shaken but were still secure.

The annihilative potential of the atomic bomb and the uneasy cold war situation of the 1950s and 1960s were important conditions for collective action organizations in many nations. The developments prompted many people to think globally and activate themselves in non-class and postmaterialist movements. The first Swedes to express postnuclear opinions did so in rather disorganized fashion and used methods of protest reminiscent of those from the first period. They opposed the proposal for a Swedish nuclear weapons program, an issue hotly debated by the military and public in the late 1950s. Many of the participators were female and young. Opposition here mobilized along three lines: (1) critical voices came from established political parties and collective action organizations, and this at times led to internal disagreement on the issue; (2) older peace movement organizations became involved, and (3) citizens opposing the bomb began to view established Swedish collective action organizations as inadequate vehicles for policy change. This last group searched for other organizational platforms, which led to the formation in 1958 of a new collective action organization, the Ad Hoc Group Against Swedish Atomic Bomb (AMSA).

The AMSA was created by several well-known Swedish intellectuals and the quakers. Some established social movements were involved as well. Authors, artists, academics, and even the Swedish archbishop– but no natural scientists, politicians, or MPs– were members. The AMSA was heavily attacked by the Social Democratic Party, but the party itself was divided on the issue. Most members were about fifty years of age. It was organized loosely, which makes it difficult to determine exactly the number of members or supporters; an estimation is 20,000 (Elmbrant, 1964). Its membership represented a cross-section of Sweden. Its aim was to mobilize public opinion through meetings and petitions, and it received a good deal of attention from the mass media.

The AMSA lost its impetus when the social democrats, liberals, and Center Party decided against a Swedish atomic weapons program in 1959 and died before it reached the life cycle stage of institutionalization. This is a good example of how Swedish collective action organizations can impact the political order and how the incorporation of ideas by the political order can effect their future existence.

A more successful attempt to mobilize support against atomic bombs was the Campaign Against Nuclear Weapons, which emerged in 1961 to inform the public about the dangers of nuclear weapons through marches and slogans. The British Campaign for Nuclear Disarmament (CND) was its organizational prototype. It also had a decentralized, loose organizational structure. Time was successfully invested in mobilizing young Swedes, and the movement began to branch out into other organizations. The Committee for Non-Violent Actions is one example. Short-term ad hoc campaigns were started and the peace, religious, and youth movements mobilized on the issue. The Committee concentrated its efforts on forming public opinion by creating unconventional events– sit-ins, demonstrations, and illegalities. It did not have much success in provoking the law and had a good, practical working relationship with the police. Relations between new politics movements and the Swedish police would change drastically in the next few years.

Initially the SFSF was very skeptical about the Committee for it did not believe that unconventional methods fit Swedish political culture. Also, it disliked the fact that the Committee allowed communists as members, an attitude held by many established collective action organizations in these years. The nuclear weapons issue helped renew the SFSF. It became a 'born-again' peace movement, with what for the SFSF must be seen as a booming and an enthusiastic membership. The SFSF experienced its sixth peak in members between 1958 and 1963.

The Swedish anti-nuclear weapons movement folded shortly after the Partial Test Ban Treaty was signed in 1963. Movement organizations had successfully brought the issue to public attention and even encouraged established collective action organizations to take a public stand. Bans on nuclear weapons became the goal of many different established collective action organizations in Sweden. One of its central effects was to change national attitudes toward citizen participation. No longer were international affairs solely the domain of an elite group of men who headed governments. Elite democratic thought was challenged by ideas about participatory democracy. Citizens were demanding a say in foreign policymaking.

From the mid-1960s concern about foreign affairs focused on the Vietnam War. The first people to raise the issue were Swedish novelists and poets, several of whom had been active in the anti-nuclear weapons movement. Then academics and journalists became involved. They were followed by young people, several of Sweden's established collective action organizations, and later government. Similar groups of writers, academics, and journalists had been instrumental in putting issues on the political agenda in the first period. Mobilization on this issue increased public interest in the peace movement, which experienced its seventh peak in membership between 1965 and 1975.

The anti-Vietnam movement in Sweden was one of the first and largest in Europe. It involved many teenagers and students without experience from politics and collective action organizations. What united the various groups in the movement was commitment to peace in Vietnam. Some of the more radical groups supported the Vietnamese Front for National Liberation (FNL, 'Viet Cong') and generally desired to fight what they considered U.S. imperialism. There was a pervasive anti-U.S. attitude in Sweden in these years, which contrasted drastically with the older view of the United States as a country of opportunity, progress, and freedom.

The established Swedish political order was drawn into the struggle. Many of the young participants were politically non-conformist and anti-establishment. They wanted to change the cultural and political landscape of

Sweden and carve out a public space respectful of young people. The anti-Vietnam and student movements were among the few organizations in the early 1960s that accepted and channeled this sentiment. Most if not all influential political parties and collective action organizations supported the established political order in the 1960s, were intermeshed with Swedish corporatism, and wanted to recruit young people but were reluctant to let them play an influential role in organizational politics. Students, like miners, are an important group in most protest cycles.

Confrontation with the Swedish establishment and political opportunity structure occurred in various ways. The youth-run and most radical and militant anti-Vietnam movement organization, the United FNL Groups (DFFG) challenged traditional political wisdom by accusing the Swedish labor movement and government of being a U.S. ally rather than a representative of an alliance-free nation. The organization was very sympathetic to new left ideologies, particularly Maoism, and later played an important role in the formation of a new left political party. The DFFG functioned as a loose network organization for local FNL groups (see Tängerstad, 1992/93).

It rejected conventional means of political communication and appealed to government through demonstrations, created events, and street uprisings. Its character was anti-authoritarian, and it used shock tactics to create public opinion and relied on means reminiscent of Sweden's first collective action organizations. Supporters sold handouts and newspapers on street corners, and outside such well-visited stores as the state liquer store, which is called 'systembolag' in Swedish. (A book on the new left movement in the 1960s has the well-chosen title, 'Outside the System' (*Utanför systemet*) (Engberg et al., 1978). The organizational newspaper had at most a circulation of 140,000 (Möller, 1992, p. 19). It held study circles which not only gave supporters the opportunity to discuss the war issue and organizational cause but helped them develop their own individual political resources. Many DFFG activists became prominent civic leaders in the 1970s and 1980s. The DFFG supported itself with voluntary donations– not government patronage. Sympathizers contributed money and time generously to the cause. A Swedish poet called it 'the most devoted, proudest, and best social movement that ever changed political opinion in this country' (Karl Vennberg quoted in Möller, 1992, p. 22, my translation).

From the start the established political order reacted very negatively. It condemned the ideology and actions of the DFFG. Initially there were a few serious and violent confrontations with the police; civil disobedience led on occasion to arrests and legal action. A working relationship with the police soon developed, and constitutional rights of demonstration were later broadened.

This book has shown that negativeness has been the initial established response to new political ideas, behaviors, and collective action organizations. Condemnation by the establishment did not hinder the mobilization potential of the DFFG. It continued to grow, and this led the established political order to adopt a new approach. More establishment-oriented anti-Vietnam movement organizations were created. The establishment of the Swedish Committee for Vietnam (SKfV) shows how Sweden's established political order integrates new ideas which it has not been able to stifle.

The SKfV was created along traditional organizational lines: a board governed it, and members came from established collective action organizations, e.g., political parties as well as the union and free church movements. Its purpose was to win over young people from the DFFG to established Swedish collective action organizations, in particular social democratic ones.

Icy relations between the two main anti-Vietnam War movement organizations began to thaw when the SKfV organized a march with the participation of the North Vietnamese ambassador. His involvement convinced the DFFG to join them. Organizers of the march believed that his involvement was a golden opportunity to mobilize support for the SKfV, but the presence of the Swedish Minister of Education Olof Palme at his side led to a freeze in diplomatic relations between Sweden and the U.S.

The Swedish anti-Vietnam war movement declined after the 1973 Paris agreement on cease fire and the U.S. decision to pull its troops out of Vietnam. The fall of Saigon in 1975 terminated the movement. It proved difficult for the DFFG to keep up the sense of urgency for the cause among its members. Protest had been a life style for its supporters. They had sacrificed their private lives and quite understandably experienced a sense of exhaustion and over-commitment. Unsuccessful attempts were made to renew organizational goals. It was easier for the SKfV to survive and renew its goals because of its backing from established organizations. It changed its name to the Swedish Committee for Vietnam, Laos, and Cambodia and continues to exist today.

Critique of the established political order was then channeled in new left parties in the late 1960s, none of which have had any significant political impact. Yet activism in them created a 'political correctness' castigating everything bourgeois and framing all discussions in terms of new left political dogma. Demise of the Swedish FNL movement had a negative impact on new left parties, whose members began to turn their attention to the more urgent issues of nuclear power and ecology. Some new left dissatisfaction was channeled to the counter-culture movement, which encouraged people frustrated with established politics and society to choose a different life style and future. Anti-commercialism and non-conventional modes of behavior are examples of its concerns.

Frequently counter-culture movement organizations functioned as loose networks for rather fluid gatherings of sympathetic groups whose aim was public opinion formation. The movement itself never mobilized a great deal of support, though mass media coverage tended to be good if not always positive and new left parties were skeptical of it. Most sympathizers were rather young. Some of the groups had a local focus, which makes them early NIMBY organizations.

Locational NIMBY actions showed that Swedes challenged the directives sent down by central government. Unconventional tactics such as occupations, sit-ins, and tree hugging were used to prohibit official action. Struggles occurred between local populations and central government over exploration drilling for final storage centers for spent nuclear fuel in various areas and over river damming for hydroelectricity. A national association for river conservation was created in 1974.

The counter-culture movement is important because it embodied the critical radicalism of the 1960s, which encouraged people to challenge traditional authority and value systems. The effects trickled down to many parts of society and even to the girl and boy scout movement, which completely deleted one of its tenets from 1936– 'A scout obeys his or her parents, teachers, and leaders'– from the 1970 scout code of honor (Sidebäck, 1992, p. 110, my translation). Participatory democracy was taking hold in Swedish political culture.

Postwar and counter-culture movements encouraged citizens to take greater interest in political and social problems. They were engaged in a general awakening of Swedish citizens and played the same general role as the free church, labor, and teetotaling movements of the first period. Also, they furthered change from material to postmaterial values. The counter-culture movement branched out into other areas– farming, product imports, and production– and now lives a rather anonymous life in Sweden. It has not reached the life cycle phase of institutionalization.

Young people were not the only group to become a political factor in the 1960s. Women were also awakening. They too wanted a say in government. Their issues became political agenda items for the second time in civil society history. A new women's movement grew out of the turmoil of the 1960s, and older women's organizations were rejuvenated. In 1968 dissatisfied young, new left women formed an organization called Group 8, which adopted a socialist perspective that considered gender discrimination and class oppression to be related and inherent problems of capitalism.

In these years, a loose national network of militant women's liberation organizations from various Swedish cities encouraged women to act collectively as women for women's rights. A new collective identity for women was forming. The militancy of the new branch of the women's movement provoked many people and led to a general public awakening of women's rights in society and politics. Women were encouraged to be less loyal and humble and use direct action– assertive, open confrontation of discrimination– to fight for their rights. A significant event in many countries was mobilization of support for free abortion rights (Eduards, 1992).

The Swedish establishment heard the message of women liberationists loud and clear: the social democratic women's caucus became more assertive; the FBF adopted a more militant image; political parties began to stress gender equality, and parliament passed legislation to improve the situation of women in society. Liberation of women was debated publicly, and movement organizations of the 1960s impacted politics by broadening the boundary of the political to include issues of gender. Success for the movement here did not mean that the establishment opened its decisionmaking arenas to women.

The new women's movement reached its zenith between 1970 and 1977. It received considerable coverage in the mass media and engaged many women in the struggle for liberation but never became institutionalized as a social movement. Many organizations declined, and the most radical of them, the Group 8, exists today in the form of an editorial board of a women's magazine. Women continued to act collectively, but their sense of urgency and priority changed. A gradual transformation from 'a feminist socialist perspective to a feminist perspective based upon a politization of a so-called women's culture of nurturance and care' took place (Peterson, forthcoming, p. 11). Feminist issues include world peace and violence against women.

The issue of nuclear power created the most turbulence in these years. It led to the clearest confrontation between material and postmaterial values ever experienced in Sweden. The issue was not controversial before the 1970s, though a small, rather unknown group against nuclear power existed and a well-known environmental activist publicized his views against nuclear power in the late 1960s.

General political consensus existed early in this period on the benefits of production of nuclear power to ease Swedish dependence on imported fuels. Even conservationists saw nuclear power as solving problem with river damming for hydroelectric production. Sweden had invested heavily in nuclear power plant construction in the 1960s, and all major decisions on the issue had been supported by the five parliamentary parties. All this changed in the early 1970s, when two women MPs forced the Center Party to question the political and environmental sensibilities of nuclear power. Its stance against nuclear power led to good results in the 1973 parliamentary election.

Sweden had become fertile ground for environmental consciousness on the dangers of nuclear power. The background of the rise of the issue to national prominence were the debates in the 1960s on pesticides and quick silver. Rachael Carson's book *Silent Spring* was translated into Swedish. Locational environmental NIMBY conflicts had begun, and the Swedish conservation movement was transformed during these years from a 'herbarium' to an 'environmental warrior' and from 'merry naturalists' to 'saviours of the nation' (Lindfeldt, 1979, Söderqvist, 1986). Authors of a comparative study of environmental movements conclude that Sweden went through three stages of green consciousness in this period: environmental awakening or sense of urgency (1960-67); ecology (1968-73), and role as an anti-nuclear movement (1973-80) (Jamison, Eyerman, and Cramer, 1990).

Environmental movement organizations were involved in the nuclear power debate, as illustrated by the opinion formation activities of two network organizations for the movement, MIGRI (established in 1971 and later called Environmental Guardians) and Action Stop Nuclear Power (1974), which included environmental, peace, and women's movement organizations as well as local units of established political parties. The base of the anti-nuclear movement was broad and strong. By 1975 the Left Party Communists had joined the anti-nuclear camp.

Increased public awareness of the dangers of nuclear power created problems for the established political parties. A broad anti-nuclear power movement began to take form in Sweden in 1978. The network organization took the name the People's Campaign Against Atomic Power, which later changed to the People's Campaign Against Nuclear Power. After a motion calling for a general referendum was voted down in parliament, anti-nuclear movement organizations and parties began in the spring of 1979 to collect over 520,000 names for a petition calling for a referendum. The Three Mile Island nuclear plant incident convinced the other parties that a referendum was necessary. Nuclear power was simply too hot for old politics to handle.

Voters were given three different choices in the referendum, but all of them called for a phasing out of nuclear power. The issue was how soon it would be implemented. The constellation of collective action organizations behind each choice shows how the campaign was structured. Organized business and the Conservative Party favored the first pro-nuclear alternative; labor movement organizations and the Liberal Party mobilized support for a second pro-nuclear alternative which boasted that it would phase out nuclear power in a commonsensical way, and several environmental, free church, and peace movement organizations as well as the Center and Christian Democratic Parties and Left Party Communists gave their support to the third alternative, which proposed a quick phase-out of nuclear power.

The labor movement stated officially and publicly that it would not cooperate with organized business and the Conservative Party for traditional, ideological reasons. This was their explanation for the decision to support the second pro-nuclear alternative. Yet what most people did not know at the time and still do not know today is that there was secret cooperation between organized business and labor in the referendum campaign (IF, 1980). The decision was actually an attempt by the labor movement to introduce old politics into the referendum campaign so that the relatively large group of social democrats opposing nuclear power would not vote for the third alternative. Postmaterial thinking about nuclear power upset the traditional left-right continuum of Swedish politics in other ways as well. The third alternative represented a very heterogeneous grouping of people. Different organizational cultures– farmers, Pentecostalists, communists, and anarchists– were forged together in the campaign. In fact, it involved more collective action organizations than the other two alternatives.

Sweden's social movements mobilized on the issue, and they used both classical Swedish and new Norwegian social movement strategies. The campaign was modeled after the anti-EEC campaign in Norway. Large marches– one from 1977 involving 20,000 people– were organized, civil disobedience was used in the beginning, and emphasis was put on mobilizing all kinds of resources. Anti-nuclear groups within the TCO and LO as well as the Liberal and Social Democratic Parties worked for the third referendum alternative.

Alternative 1 received 18.8 percent, Alternative 2 39.2 percent, and Alternative 3 38.7 percent of the vote. More younger and old– but not middle-aged– people and women voted for Alternative 3. This made the issue an exception to the general rule of class voting in Swedish elections as gender and age were important determinants of referendum choice (Holmberg and Asp, 1984). Supporters of Alternative 3 were disappointed by the referendum results, but what disturbed them most was the decision by organized business and labor to pool their votes and declare themselves the winners. Considering the campaign rhetoric used by the Alternative 2 camp, a very conceivable choice would have been to combine Alternatives 2 and 3.

Many sympathizers had devoted themselves entirely to the campaign and considered this move to be a manipulation of democratic politics. This led to disillusionment with democratic politics (Eriksson et al., 1982, Sjöberg, 1988, Holmberg and Asp, 1984) and a shift from collective to more private involvement for many of them. Such involvement included participation in the new age movement, whose focus is individualism and moral strength, and commitment to a more home-grown new ecologism that stressed nature preservation. Certain sympathizers turned their attention to the peace movement, which mobilized support for a nuclear free zone in the Nordic region and experienced its eight mobilization peak between 1979 and 1984.

Disappointment here is parallel with the feelings of the teetotaling movement after its defeat in the 1922 referendum on prohibition. The other main similarities between these two general referendums are the important role played by collective action organizations, gender dimension of voting, and careful phrasing of the referendum alternatives which promoted status-quo policy. The major dissimilarity is that the anti-nuclear power movement did not make peace with the political order by adapting its goals to government policy, though certain leading activists became ecocrats, i.e., environmental activists working in the public administration. The anti-nuclear power movement renewed its dedication, broadened its goals to include protest against nuclear weapons, and changed its name to the People's Campaign Against Nuclear Power and Nuclear Weapons. Today the movement exists as a semi-institutional, low-keyed environmental and peace movement organization.

The outcome of the referendum came as a blow to the environmental movement. Its leaders reflected seriously upon the role that the movement had played in the referendum campaign. Questions were raised about organizational structure and strategy as well as the close bonds with established political actors, which led to movement dependence on the political order. Many leaders believed that loose, flat networks as the People's Campaign were ill-suited for the Swedish political opportunity structure. Gradually a more cynical view of Swedish political democracy transformed the movement. Grassroot strategies were replaced by environmental management, media politics, and the Greenpeace model of support mobilization. Newer, smaller, more militant movement organizations took form, and an environmental party was established in the next period.

Nuclear power had dominated public debate and the mass media for a decade, a unique situation according to election researchers (Holmberg and Asp, 1984). The referendum results showed that postmaterial values were not fully developed among Swedish voters, and Swedish politics lacked a clear green dimension. Yet the debate on nuclear power put postmaterial issues on the political agenda, and they have remained there ever since.

Impact of change in Sweden's political opportunity structure

Uncertainty about Sweden's future was not only felt by sympathizers of the new collective action organizations of the 1960s. Established political actors began in the 1950s to debate whether the constitution that had governed Sweden since 1809 was still a living document. They saw problems with the parliamentary system, electoral laws, and general framework of political rights guaranteed to citizens. Voting age as well as suffrage for immigrants were important issues.

One of the basic problems was the bicameral parliament. The Upper Chamber with its eight year term of office and indirect election procedures favored larger parties and preserved the influence of parties with declining electoral support. Swedish citizens directly elected MPs to the Lower Chamber every four years. This dual system caused problems. Theoretically parliament could have been deadlocked and Sweden ungovernable were it not for ad hoc solutions. The non-socialist parties argued that the system conserved the political power of the social democrats. The desirability of replacing bicameralism with unicameralism and reforming the constitution were issues high on the political agenda in the 1960s.

It is unfortunate that the general debate on constitutional reform is not one of the high points in Swedish political history. The political parties bickered with each other; debate on constitutional principles was overshadowed by concern for vote maximization, and a commission appointed to investigate constitutional reform in 1954 failed to offer a convincing proposal in its final report from 1963 (von Sydow, 1989). Yet after years of investigation and debate, a unicameral parliament was established in 1971 and a new constitution adopted in 1974. Voting age was gradually lower to 18 years, and immigrants were given the right to vote in local and regional elections. Several aspects of this change in political opportunity structure are important for Swedish collective action organizations.

It is worthy to note that collective action organizations played a minor role in the constitutional debate. All peak labor market organizations, farmers' movement organizations, the IF, KF, and Swedish Bar Association were sent the final reports of the government commission for comment. Certain other organizations were asked to submit comments on special parts, among them the FBF, National Union of the Swedish Saami People, Swedish Retail Federation, and Federation of Swedish Commerce and Trade. The KF did not submit any reply, and several others restricted their comments to specific issues dealing specifically with the special interests of their members.

The collective action organizations did not express an opinion on unicameralism. Labor market organizations are usually very active in the commenting process but were rather passive this time around. This seems strange given the effects of the general political opportunity structure on the function of collective action organizations, but they most likely felt that constitutional change was the responsibility of political parties and the government. The only real exception here is their comments on their own role in the Swedish political process. The government commission on constitutional reform appointed a special study of this (Foyer, 1961); the commission approved the role that organizations came to play in Swedish politics and proposed that the new constitution (Instrument of Government) give them mention as legitimate actors in the commenting process. The commission did explicitly caution against the development of corporatism, which it defined as legitimation of certain special interests as legislative representatives of social groups (SOU, 1963, no. 17). A second state commission appointed to prepare constitutional reform reiterated these standpoints (SOU, 1972, no. 15).

Constitutional change impacted Swedish politics in numerous ways. It strengthened the democratic legitimacy of parliament which, among other things, meant stricter proportional representation and government more directly dependent on parliament for its formation and survival. The balance of power in the Swedish political system shifted towards the legislative branch, which increasingly took political initiative and a independent position in politics (see Isberg, 1982). Years later it became common to speak of the reparliamentization of Swedish politics. Parliament even became more accessible for the public. Committees held a greater number of closed and later open hearings, and certain collective action organizations began increasingly to lobby parliament. The non-socialist governments (1976-82) encouraged lobbying because they generally condemned corporatist intermediation and promoted pluralist methods of influence. The SAF was the first collective action organization to professionalize its efforts by creating a lobby unit at its headquarters (Micheletti, 1984a).

The three-year parliamentary term of office meant that politicians had to work faster to fulfill campaign promises. This led to a new conception of political time, and politicians, journalists, and political scientists began to speak about the importance of a 'flying start' for new governments. This development put more focus on political accountability of parties and the government to the electorate. Collective action organizations had a good sense of the impact of this change, and the first primitive interest group rating of the parliamentary parties since the early years of Swedish civil society was made by a small, non-institutionalized environmental movement organization, the Environmental Federation (established in 1976) during the 1979 election campaign. Interest group ratings and involvement of collective action organizations in electoral campaigns would later become a commonplace.

The problem for the political system was that the effects of constitutional reform were magnified by new volatility among Swedish voters. Some parliamentary parties– for instance the Center Party– experienced rather strong fluctuations in support in the 1970s. Also, established political parties were forced by changing employment and demographic patterns to mobilize support from new groups, and they found that their traditional voter groups were not only shrinking in number but much more heterogeneous than in the past. New political parties on the left and center formed. The most important one is the Christian Democratic Party (KDS), founded in 1964, which mobilized supporters from the Pentecostal movement as well as groups that disagreed with the general secularization of Sweden. The KDS competed with the Center and Liberal Parties for votes and became the largest of the nonparliamentary parties in Sweden.

Change in political opportunity structure definitely impacted Swedish politics and civil society. There is a clear break with the past. Political security and stable contacts between government and established collective action organizations characterized the periods of the strong society and Swedish model. The 1960s experienced the struggle to democratize politics nationally, locally, and organizationally. Concern in the 1970s involved governing the polity in trying political and economic times (cf. Tarschys, 1978, Weaver, 1987).

The public was made more aware of the governability crisis by the opinion formation apparatus of organized business, which published many book on the general topics of government overload, overregulation, and sluggishness. The SAF began actively to oppose Swedish corporatism towards the end of this period. It was characteristic of the times that collective action organizations and particularly labor unions were seen more as a cause than a solution of government problems. The political right was mobilizing itself.

The Swedish tradition of granting large, established collective action organizations access to all stages of political decisionmaking was seen as an important cause of Sweden's economic and political crisis. The practice of corporatism was not only questioned but highly debated in the late 1970s. A common opinion was that it gave members of collective action organizations extra suffrage rights. Many leading non-socialists believed that corporatism and social democratic hegemony were reenforcing.

All political actors agreed that labor union organizations had grown strong by this period of civil society history but disagreed about whether this was a good or bad development. Unions organized more members than ever before, and their spheres of activity became much broader. Legislation from the mid-1970s improved industrial democracy greatly, and the Co-determination Act (MBL) of 1977 even made it possible for public sector unions to exercise influence over government bills and decisionmaking. Unions used extreme caution in exercising this right and negotiated for only marginal changes in planned political decision. Yet the non-socialist parties adamantly criticized its mere existence, and the law was later modified.

In many ways, it became difficult to distinguish between large, established collective action organizations and political parties in this period of dissatisfaction, political turmoil, and reparliamentization of politics. Organizations became part of the public and the public part of them. They became more similar to political parties in that they were also forced to mobilize support among the public in order to attain their goals. No longer were contacts with the Prime Minister and cabinet heads sufficient to ensure satisfactory political outcomes. What the public thought about the goals of collective action became a crucial element in organizational planning. Successful attainment of goals by large established collective action organizations increasingly meant a financial or policy commitment by government. The crux of the matter was that even the public had to support their goals. It became particularly important to mobilize broad public support. Stable alliances with one party could no longer guarantee satisfactory political outcomes, and minority parliamentarianism and electorate volatility forced collective action organizations to find new ways to influence politicians. Their answer to this change in political opportunity structure was public opinion formation (Micheletti, 1985a).

Organized business took the lead. The SAF and IF contributed financially to the Free Enterprise News Agency, a research center called Business and Industry Information Group, and two publishing houses. The public relations apparatus of organized business was very active in the wage earner fund debate as organizational leaders believed that the non-socialist parties were slow to react to the proposal. Also, they were disappointed that the non-socialist governments did not delimit the welfare state. They decided that their best strategy was to mobilize direct support from the public (Micheletti, 1984a, De Geer, 1989).

The strengthened role of the mass media also spurred on public opinion formation. Journalists became more active seekers and interpreters of information. Collective action organizations found, for instance, that their democratic practices were scrutinized more carefully and their press releases and public information no longer automatically made news. These developments forced organizations to adopt more conscious and offensive media strategies. They began to compete with political parties and other collective action organizations for media attention (Micheletti, 1985a).

Collective action organizations were impacted by two other changes in Sweden's political opportunity structure, municipal mergers and decentralization of national authority to regional and local government units. Municipal mergers took place in three waves and decreased the number of local government units from about 2,500 in the early 1950s to 280 by 1974. The need for dynamic municipalities that could run welfare services effectively was the main reason for reform. Mergers were not supported by all citizens, and local anti-merger groups formed in certain municipalities; some were even successful in impeding implementation (see Wallin, 1991). This transformation in the size and democracy of local government changed politics in important ways, and citizens compensated the greater distance between themselves and their politicians by using collective action organizations to channel their interests and represent them in contacts with local officials. Local collective action organizations began to play a more political than social welfare role from this period of civil society history. They became more active in municipal politics.

Larger organizations were most affected by the mergers, while smaller ones continued to work within the same limited radius. Most organizations adapted their activities to the new boundaries; very few were active in more than one municipality. Studies on the impact of this change in political opportunity structure on collective action organizations found that organizations critical of established local political order tend to exist in areas characterized by diversity of opinion, i.e., larger, urban municipalities (Gidlund et al., 1982, Bergqvist, 1969). This explains the emergence of new social movements in large Swedish metropolitan areas in the 1960s and 1970s, but it should be noted that many NIMBY groups have been active in sparsely-populated areas as well.

Collective action organizations are involved in different ways in local politics. Local corporatism developed in Sweden. Most municipalities had regular contact with collective action organizations, and a good number of local politicians are also active members of organizations. Many municipalities regularized contact by holding conferences with all local organizations or by employing liaisons associated with different local sectoral boards. Organizations give comments on municipal proposals, and attempts were made to integrate them in municipal planning. Local government and collective action organizations have joined efforts to reach specific groups of citizens by establishing formal advisory councils for handicapped people, senior citizens, and immigrants.

Decentralization of state authority to local and regional authority began in the late 1970s and continued in the next period. Its purpose is to alleviate problems caused by local government mergers and from policy sectorization and centralized control of welfare services. Most importantly, decentralization is seen as a way to bring government closer to the Swedish people. Collective action organizations did not really respond seriously to this important change in Sweden's political opportunity structure. Some of the most established ones and particularly the peak labor market organizations further enlarged their central operations by localizing tasks to the local and regional level. Organizational decentralization took place in the next period of civil society history.

Towards setting a boundary between the Swedish state and civil society

Several of Sweden's most glorified collective action organizations experienced serious difficulty with adapting themselves to the new political opportunity structure and changing times that characterized this period. The problem attracted public attention after the ABF and social democratic youth movement called a series of conferences on the future role of social movements. Two central concerns of the public debate were how municipalization and the dominance of the culture of the strong society and Swedish model impacted social movements. Both developments shaped collective action organizations after government design. Popular books were written on the subject; central government decided to take action to preserve and revive Sweden's venerated social movements, and contracting out of welfare services to them was proposed.

Few Swedes find government involvement and integration with civil society as strange. This is indeed a historic paradox for true social movements are independent of government and the established political order. Yet government involvement with collective action organizations shows how the movements had become goals in themselves rather than channels or methods of social protest. Central government invested in them financially, politically, and gave them carte blanche legitimacy. Government has been a main source of glorification of collective action organizations and particularly the classic Swedish social movements.

A substantial investment in collective action organizations was made in 1975, when a reference group for social movements, abbreviated here as RGSM, was formed within the Ministry of Municipalities. It is a liaison network for social movement organizations and government that aims at renewing social movements in all ways and forms. A minister responsible for government policy in this area is customarily called the minister of social movements. Among its activities, it publishes a guide to Swedish social movements which is of particular interest as it implicitly offers a governmental definition of national social movements. The reference group initiates conferences and produces reports on how social movements can renew themselves in the face of social change. Yet the major reason for its creation was discussion in government on contracting out of welfare service and the need for well-functioning and well-equipped social movements to take over the tasks.

Municipalization created a rather unclear and confusing division of responsibility between government and collective action organizations. Politicians frequently chose it as a way to cut costs. This solution did not always satisfy social movements organizations, which found themselves unable to compete with central and local government services. Concerned citizens began to wonder whether municipalization was a good idea; criticism became more pronounced in the early 1970s, and the government began to believe that 'there are many arguments for transferring tasks that social movements performed well or better from the welfare state back to social movements' (Speech by Minister of Municipalities Gustafsson, 1975, my translation). Contracting out was seen as a solution to government overload. The idea was to contract out certain tasks and give collective action organizations greater responsibility for social contact services for different categories of citizens–handicapped people, senior citizens, children and young people with free time, and immigrants. This idea gained currency as a viable alternative early in the next period.

Questioning established collective action organizations

Changes in Sweden's political landscape impacted social movements in many ways. Electoral volatility signified profound changes within social classes. New employment patterns greatly affected collective action organizations in general and the labor market organizations in particular. Class solidarity, which was the social glue that checked the development of diversity of opinion within the LO, began to break down especially among younger and urban union members . Even the Swedish class of businessmen became more diversified, and this led to a transformation of the SAF into an employer association for small business actively involved in sectors far removed from Swedish industry.

Heterogenization of organizational memberships and demands by members for more participatory democracy created problems of governability for established collective action organizations. Use of older organizational methods and insistence on retaining traditional values contrasted greatly with the spontaneity, creativity, and ability of new social movements of the period to capture the zeitgeist. Most leaders of established social movements condemned NIMBIES, counter-culture movement organizations, and all new social movements for lacking a unified ideology and long-term goals and for using unconventional methods. Yet their ability to see problems within their own organizations was limited to complaints about the paralyzing and bureaucratizing effects of government patronage, commercialization and mass mediazation of society, and difficulties of getting members involved in a discussion of movement ideas, principles, and goals.

Few leaders of established social movements understood that the general goal of new and different groups was the same as those which their own movements promoted in their genesis phase. New social movements desired to redefine politics, write a new social contract, and rethink social and economic progress. As their first wave predecessors, they are pioneers of democracy. A speaker at a conference on the future role of social movements held by the teetotaling movement came close in expressing this:

> The goals of the teetotaling movement have been discussed in recent years, and it has been stated that they are not ambitious enough– e.g., discussion on a alcohol-free society vs. a society free from the problems that alcoholism create....I myself have always believed that the primary goal of the teetotaling movement is to create a better society, to improve the situation for individual people....Today it is clear that we are faced with difficult problems. We were once protest movements...We cared for social outcasts and the poor. We were ourselves more-or-less outcasts. Today we are established members

> of the political order. Government ministers bow respectfully and acclaim that they need us, though they do not perhaps always help us in their political actions. We have become part of the established order and safeguard that which we have helped develop. This makes it difficult to capture and channel the winds that blow in actual issues, no matter whether it deals with elm trees [in central Stockholm] or a waterfall in Northern Sweden. We have become part of the political order, we socialize our members to work within this order and, of course, make it better but at the same time we generally accept it. The major dilemma for social movements today is that they have difficulty in improving society and channeling the short-sighted opinions which citizens express locally, regionally, or nationally that criticize the established order which they have together with others created and want to safeguard (*Nykterhetsrörelsens Landsförbund*, 1972, pp. 38-9, my translation).

Established collective action organizations experienced a crisis of public legitimacy. Segments of the public questioned their structure, processes, and range of activities. Elite democracy was seen as less representative of members, as poignantly revealed in the LKAB and nuclear power examples discussed earlier, and groups of underprivileged members demanded a voice in organizational affairs. Corporatist interaction and settings were seen as blocking young people, immigrants, and particularly women from exercising influence in their organizations (Hernes, 1987, Bergqvist, 1991). Corporatism led to compromises between organizations and government that had not been anchored in member opinion and interest inarticulation. The legitimacy of the distributional coalitions– the organized farmers, male workers, and businessmen– that had been the backbone of Swedish politics was breaking down rapidly. Social classes had become very fragmented (cf. Petersson, Westholm, and Blomberg, 1989), and new social identities were beginning to take form.

People holding postmaterialist values did not cause a silent revolution in Sweden but their points of view were indeed leading to a cultural shift, as expressed in the creation of fourth wave movements and their ability to mobilize significant segments of Swedish society. They included environmental, counter-culture, new religious, and global humanitarian collective action organizations. Some established collective action organizations took measures to incorporate new Swedish thinking into their programs. The adult educations associations stand out in this regard. Compared to other established organizations, they have been at the cutting edge of new individualist, counter-culture and postmaterialist thinking. For instance, the Study Association of Teetotalers (NBV) offered study circles on health food and creative drama. The Non-Conformist Churches Study Association (FS) held study circles on citizen empowerment, new individualism, and Christian renewal.

The free churches were among the first collective action organizations to reach out to immigrants, a social group in Swedish which grew numerically. The political opportunity structure for immigrants changed after government adopted a new immigrant policy that acknowledged and encouraged cultural pluralism. Perhaps the churches sensed that the new government policy would lead to ethnic clashes and a clear division between Swedes and non-Swedes and desired to prevent this by integrating immigrants in their activities. Many new religious movement organizations were rather successful in mobilizing support among immigrants, who longed for a sense of community in their new homeland. Also, several immigrant churches were founded in this and the next periods.

Competition from less or non-established religious groups helps also to explain emphasis on new individualism by the established free churches. Approximately thirty new religious movement organizations, many of which were New Age and non-Christian in orientation, emerged and less conventional religious affiliations such as Jehovas Witness gained support. Finally, more radical identities were forming around gender, age, and ethnicity. Young people and pensioners began to demand more say in politics. Senior citizens mobilized in their own organizations would become an even more important political actor in the next period. A question which could have been asked at the time was whether these developments were generally steering Sweden down a path towards political disintegration.

7 Repluralization and resurgence of Swedish civil society

The period of the dissatisfied generation was the beginning of more permanent transition of the Swedish political system. People began to lose faith in the ability of political leaders and spokespeople for certain organized interests. The protests of the 1960s and 1970s showed that a significant segment of the population wanted politics to change direction and democracy to be expanded. The economic difficulties of the 1970s shook public confidence in the ability of government. Leading economists asked: how much politics can the Swedish economy tolerate ? Political appointees and civil servants from the Ministry of Finance in the 1980s were among the first group of government officials to question the role of collective action organizations in the Swedish political process. Later other groups did so as well. Economic decline shattered the general consensus upon which confidence in corporatism and social democratic rule rested. Low economic growth, high unemployment, inflation, and unfavorable trade balances created serious problems for Swedish policymakers. Swedish corporatism was criticized by all political parties, including the Social Democratic and Center Parties whose electoral constituencies constituted two of the three pillars upon which the strong society and Swedish model rested.

This period (1980s to 1990s) is also characterized by sensitivity to and awareness of aspects of life that mainstream politics tended to ignore. New political issues reached the public agenda. Environmental accidents made Swedes acutely aware of their global dependence. The Environmental Green Party was the first new party to enter parliament since the early 1920s, when the Agrarian Party was voted into the Lower Chamber. Feminism began to gain serious ground in politics. Issues of ethnicity became political agenda items. The world's refugee problem became more serious, and anti-immigration attitudes found organized form in new right political parties and racist organizations.

The entrance of the KDS in parliament after the 1985 election and government in 1991 and the mobilizing potential of Christian fundamentalism showed that the country was not that secularized. New religious movements tend to appear in times of transition and strife. The first social movements to emerge were religious, and they catered to the needs of Swedes who felt rootless in the transition of the time. Swedes began searching for new ways to interpret their individual and national identities; new religious movements were one alternative open to them. The eventuality of Swedish membership in the European Union (EU) and the process of internationalization have put the issue of national identity high on the political agenda.

The emergence of new social movements and political parties shows that established Swedish parties and collective action organizations are losing their ability to mobilize mass public support and that they cannot cope with internal struggles for power. The issues of struggle are the role of women, young people, and immigrants as well as the direction of politics. Established parties have become less successful as linkages in helping citizens and government interact with one another. Sweden is experiencing a crisis of political representation, a development she shares with other western, industrial nations. Mass society develops when linkages between citizens and government are no longer effective.

End of social democratic hegemony

A central reason for the search for new identity is the waning importance of established political parties, and particularly social democracy, as the main reference point for considerable numbers of Swedes. The 1980s marked the end of social democratic hegemony. Erosion of hegemonic control over a country does not only mean that the dominant party begins to lose its clientele and finds it difficult to reproduce political bonds across generations but that its opinions and ideology are no longer self-evident points of departure for public debate.

Studies of the parliamentary elections held reveal a general decline in partisan loyalties and an increase in issue-based voting. Also, they show that voters considered themselves to be less leftist than previously. The change in political zeitgeist has been most pronounced among young people, who represented the most leftist group in the 1979 election study. By the 1985 election Swedish voters were across the board and young people in particular more negative to social democratic politics– e.g., wage earner funds and government control of private business– and more positive to non-socialist demands for lower taxes and better defense (Holmberg and Gilljam, 1987). A similar development was noted among Swedish parliamentarians (Verba et al., 1987, Holmberg and Esaiasson, 1988). Neo-liberal attitudes had taken hold in Sweden.

Organized business professionalized and expanded its opinion formation potential in order to challenge the political wisdom of the social democrats, and collective action organizations of the new right have been actively engaged in the struggle over Sweden's political hegemony. There are different reasons for the success of these efforts. One explanation is the seriousness of the economic situation facing Sweden in the 1980s and 1990s. This forced the non-socialist governments (1976-82, 1991-94) and the social democratic government (1982-91, 1994-) to rethink economic policy and the role played by collective action organizations in it.

A second important reason is the strategic mistake of the Social Democratic Party and LO in pushing the wage earner fund issue. The issue divided the Swedish nation, polarized politics, and caused severe problems for certain established political actors. The TCO experienced a revolt by adamant non-socialist members who formed a protest group and a negative media campaign that has no parallel in the history of the white-collar union movement (Micheletti, 1985b). The issue was also the beginning of a public 'war of the roses' within the labor movement. Prominent members of the party criticized the proposal, and it became clear that a party wing was less enthusiastic about the dominant role played by the LO in internal party politics.

The wage earner fund issue was a gold mine for the non-socialists and gave the three parties the opportunity to show that they could in fact agree on a key political issue. Such agreement among the three non-socialist parties was otherwise a rarity in this and earlier periods. The issue prompted organized business into action: it created media events, put more resources into opinion formation in the 1982 election campaign than in any other campaign for decades, and held one of the largest protest demonstrations in Swedish history in 1983 which drew over 70,000 participators. The purpose of the demonstration was to publicize widespread opposition to the social democratic government bill on wage earner funds. A special ad hoc group called the Fourth of October Movement, with origins from the 1976 general election, mobilized support for and sponsored the demonstration which was held in Stockholm on the first day of the fall session of parliament in 1983.

Newspaper columnists called the demonstration a belated non-socialist revolution; partisan supporters believed that a new social movement had emerged and a turning point in Swedish politics reached (Micheletti, 1984b, Larsson, 1983). The ad hoc group folded after the parliamentary vote in spring 1992 to disband existing wage earner funds. A German scholar of European social democracy concluded that the ability of the Swedish right to unite on the issue was a crucial symbolic blow to social democratic hegemony:

> It did not matter that the [social democratic] finance minister explained over and over again that the regional funds [government bill with a watered-down wage earner fund proposal] (financed by a tax on windfall profits and limited in their acquisition of shares) were not the first step toward socialism but the first and last step toward the workers' funds and nothing else....From now on, neoliberal and neoconservative notions enjoyed a degree of attention and respectability that was incompatible with the monopoly position that the Social Democrats and LO had achieved in the late 1950s and 1960s. In the 1980s, at any rate, their definitions of political and moral goals and their interpretation of the economic situation would no longer be accepted without challenge by the centrist parties and the other unions (Scharpf, 1987, p. 113).

A divided non-socialist bloc was important for the development and success of Swedish corporatism and a prerequisite for social democratic hegemony (Katzenstein, 1985). The wage earner issue changed this and showed that there were serious lines of conflict within the labor movement which could be exploited by non-socialist sympathizers, organized business, and the mass media (Gilljam, 1988). The wage earner fund issue even convinced organized business that it could win political struggles with the help of sophisticated opinion formation campaigns. Yet more than anything else, the issue showed that Swedish social democracy was losing its hegemonic advantage. Political truths could be questioned and political taboos discussed openly. It became possible to criticize corporatism and the strength of trade unions.

Collective action organizations were considered to be 'demand machines'. This term was coined by the leader of the Liberal Party in 1979 to describe the role of collective action organizations in contemporary Swedish politics. The characterization of collective action organizations as special interests rather than organizations which took public responsibility for their actions became more prevalent in later years of this period. Organizations reacted by becoming more defensive, and unions debated whether Sweden was experiencing an anti-union wave of sentiment.

The public debated the role, scope, and costs of government. The public administration underwent evaluation, reorganization, and transformation. Swedish pride in its welfare state faded as welfare state problems became public. Concern for effectiveness, budget costs, decentralization, and deregulation replaced faith in the strong society. Private enterprise and market competition rather than the 'people's home' became the point of departure for discussions on policy reforms.

Conception of the Swedish model began to change in the late 1970s. Several leading social democrats participated in the discussion, but a bulwark against change was the LO. This is the root of the 'war of the roses'. Conceptual revision changed the course of Swedish history and led to the downfall of social democratic hegemony. The non-socialist parties and their allies began to prepare for a sweeping systemic shift in politics early in this period, and organized business pumped money into its PR apparatus. The aim was to change completely the political landscape of Sweden. These developments were beginning to show that social democratic hegemony and corporatism were fair weather products. They succeeded when there was stable economic growth that allowed for non-zero sum solutions to welfare state reform (Micheletti, 1991a, Scharpf, 1991). This period represents a historical juncture for civil society and the Swedish political opportunity structure.

Decorporatization of Sweden

Political scientists frequently analyze legitimacy of political regimes in terms of public perception of national economic performance. National economic performance has been found to be particularly important for politics in such export-oriented small states as Sweden, which explains the development of corporatist interaction in small European states (Katzenstein, 1985). Research on the U.S. shows that economic performance explains the political influence of organized business in that 'business has tended to lose political influence when the economy was performing relatively well and has become more influential when the performance of the economy deteriorated' (Vogel, 1989, 8). The important distinction here is between control of the economy by government and political actors (i.e., labor unions) in times of economic prosperity or by market-oriented actors (companies and organized business) in times of economic decline. The U.S. experience shows that the impact of business on government is also conditioned by a nation's political opportunity structure and the ability of business to organize itself effectively. Developments of this kind in Sweden are leading to decorporatization.

When the process of decorporatization takes place in democratic states it leads either to sponsored pluralism, which implies an integrated relationship between certain collective action organizations and government (e.g., organized business and agriculture), or pressure group pluralism, in which all collective action organizations compete for political attention. Political scientists are now trying to predict the direction of the process of decorporatization in Sweden. Is it heading toward pressure group pluralism or sponsored pluralism? An analysis of the explanations for the decision by government and collective action organizations to undertake decorporatization is a necessary basis for any attempt to answer this question. This section discusses seven major explanations for decorporatization.

1. *Breakdown of Power Structure of the Swedish Model.* Corporatism works well for political decisionmaking when economic blocs are easily distinguishable and form the basis for politics. This was the case in Sweden in earlier periods, when blue-collar workers, farmers, and business formed the pillars of Swedish society and the bases of the main political parties. The constellation of political forces began to break down in the latter part of the period of the Swedish model and dissolve more rapidly in the two final periods taken up in this book. Now it is difficult to clearly discern a socialist and non-socialist bloc. An ongoing heterogenization of the memberships of several of Sweden's important collective action organizations, among them the LO, TCO, LRF, and SAF has taken place as well. The result is increased internal disagreement and difficulty in reaching unanimous decisions (Micheletti, 1988a, 1990a). This is affecting the balance between government and civil society, as depicted in Figure 1.4 in Chapter 1.

2. *Decline of Consensus.* Corporatism and consensual democracy are complementary. Consensual democracy is a way of reaching political decisions on the basis of agreement on the chosen solution by all or most actors. This contrasts greatly with decisionmaking based on numerical strength, where a majority overpowers a minority. Majority democracy and pluralism complement each other (Lijphart and Crepaz, 1991). Political decisionmaking based on consensus among the main political actors began to break down. Not only do the major parliamentary political parties hold different opinions on many major issues today. Even the four parties that formed the non-socialist coalition government from 1991 could not agree on many important political issues.

An early indication of the malfunctioning of the consensual style came in the 1970s. Criticism was directed against the rather secretive deliberations within government commissions, one of the main bulwarks of the Swedish political culture of consensus, and it was proposed that they make public their documents and alternative proposals. Consensus politics was later declared to be expendable by the social democratic government when its Office of the Prime Minister advised commissions to report their findings whether full agreement had been reached or not. The office maintained that many political issues require a public airing of alternative proposals raised in commissions. Its report marks a clear change in policy from the non-socialist governments of the late 1970s and earlier social democratic governments as well (Micheletti, 1985b). The level of disagreement within government commissions increased over the years. The system of commissions underwent other changes, all of which point to a declining role for decisionmaking by consensus in Swedish politics. This development has impacted social democratic hegemony and corporatism negatively as political problems and criticism are aired publicly. The Swedish political cultural trait of conflict avoidance is breaking down.

3. *Problems of Governability*. Government views involvement of collective action organizations in political decisionmaking as valuable when it contributes to the achievement of public policy goals and creates favorable conditions for macroeconomic policy coordination (Scharpf, 1987). Political actors began to reevaluate Swedish corporatism in the latter part of the previous period. Politicians experienced serious problems in reaching agreements when collective action organizations and particularly encompassing labor market organizations are involved in government commissions (J. Johansson, 1992). The frustration of politicians here explains the absence of several of Sweden's largest collective action organizations from commissions that worked on issues of policy reform of importance to the organizations. Examples include tax and agricultural policy as well as reform of public agency government.

Contact between government and collective action organizations has been both extensive and intensive in Sweden. A study from 1988 of the Swedish political power network showed that 67 percent of all ministers in the social democratic government reported weekly contact with at least one collective action organization, and 13 percent had contact on a daily basis. The LO had the most contact and was followed closely by the TCO and SACO/SR, with the SAF and IF not far behind. It is worthy to note that a ministry with very limited contact was the Ministry of Finance (Petersson, 1989, see Micheletti, 1993). Social democratic and non-socialist government officials have lamented the interference of collective action organizations in attempts to reform the welfare state. The outcry with the most impact came from the Ministry of Finance in the late 1980s in confrontations with the peak unions and LRF. Disagreement with the LO was an important reason for the decision of a very popular and respected social democratic Minister of Finance to leave office. Later he recounted his experience in a book that became a media event before, during, and after the 1991 parliamentary election (see Feldt, 1991).

4. *Accountability*. Problems of public accountability caused by integration of collective action organizations with government were given more political importance (cf. Birkinshaw, 1988). Questions were raised about who really was in charge of public agencies and responsible for their decisions (Marcusson and Petersson, 1990). Accountability was a problem for collective action organizations as well. Corporatism created internal problems with accountability to members. At times organizations experienced that they were hostages of government when they felt forced to sanction political decisions contrary to their own organizational ideologies. The SAF is a case in point (Rothstein, 1989). Collective action organizations found on other occasions

that corporatism made it difficult for them to know how they should act; this led to interest inarticulation (Micheletti, 1990a).

Accountability problems were hotly debated in the early 1980s after a government commission issued a report proposing that interest organizations be removed as members of lay boards that govern many public agencies. The role that collective action organizations play in the Swedish political process was directly questioned. The commission wondered who the board members from collective action organizations really represented, and this led to an intense public debate on the ability, advantages, and disadvantages of collective action organizations to assume public responsibility. The report was send out for comment. Many collective action organizations submitted highly critical comments and defended their presence on agency boards with all the vehemence that written words can muster. The TCO was especially prolific. The SAF and the SACO/SR reacted in a completely different way, for they believed that corporatist integration was detrimental for civil society and government.

Discussion here prompted the SAF to begin an expensive and ambitious public campaign to decorporatize Sweden, and it was not discouraged when the social democratic government and parliament rejected the proposal. It organized seminars on the issue, published reports and a book with the apt title 'Farewell to Corporatism' (SAF, 1991a, my translation), pressured government to change its policy, and finally decided to leave a number of the public agencies, most of which handle labor market policy issues.

The social democratic government did not respond to the decision by the SAF to voluntarily leave the boards of twelve public agencies. But the new non-socialist government was quick to react to this chance to change the political opportunity structure. It saw the decision as creating a minor constitutional crisis because the directives governing a total of twenty one public agencies directly specified that certain labor market organizations be included in board decisions. It believed that the decision by the SAF disrupted the balance of interests, which was the main reason for representation of countervailing powers in the first place, and sent a memorandum on the matter to several collective action organizations and public institutions for comment in spring 1992. It then submitted a bill to parliament that proposed, as a first step, that the remaining interest organizations be removed from the boards that the SAF had left. Its second step involved appointing a government commission to consider how public agencies should be governed. No collective action organizations are commission members.

The government argued that a new evaluation was prompted by changes which impacted how Sweden was to be ruled, e.g., decentralization of political and economic responsibility directly to the agencies. An important guideline that the government gave the commission concerned the participation of collective action organizations in advisory groups associated with the agency but not in the agency board itself. This had also been proposed by the commission from the 1980s. The replies, with the exception of the LO and TCO, generally agreed. Both unions considered it wrong to give the decision of the SAF such political magnitude as to necessitate a constitutional revamping of representation by collective action organizations on public agency boards.

No one argued that the influence of collective action organizations over public agencies should be cut off completely. The SAF emphasized in its reply that it still desires close ties with public agencies but without central government regulation. Rather, individual agencies and organizations should take contact when they find it necessary. Quite interestingly, the viewpoint of the SAF here has changed somewhat since its reply to the 1985 commission report. It argued at the time for advisory councils as a general alternative to lay board membership for collective action organizations. This implied open and stable access for those collective action organizations which traditionally have been lay board members. In the early 1990s the SAF is arguing for increased use of lobbying methods, a development which labor unions oppose for they fear that it will delimit their influence.

The parliament passed the government bill, and the debate before the vote showed that corporatism had definitely become a partisan matter. The social democrats voted against and the non-socialist parties voted for it. In the parliamentary debate, the social democrats accused the government of allowing itself to be influenced by the SAF. In the past, the non-socialist opposition criticized the social democratic government of listening too closely to the LO.

Over the years, the SAF has prepared itself well for the repluralization of Sweden by strengthening its resources for lobbying and opinion formation. Labor unions believe that the resources of organized business will overwhelm Sweden and lead to the further decline of unionism. Yet many collective action organizations began to prepare for decorporatization by stocking their arsenal of organizational weapons with lobbying and opinion formation devices even before the 1992 parliamentary decision (see Närling, 1993, Micheletti, 1985a).

5. *New Economic Thinking*. Change in economic thinking began under social democratic rule and was necessitated by the rather severe economic situation of Sweden in the 1980s. It involved replacing Keynesian full-employment policy with supply-side economic policy. This decision meant that it was no longer necessary for unions to be integrated into anti-inflationary wage policy schemes. The consequence was that national economic policy no longer needs corporatist relations between government, unions, and employer associations in the form of incomes policies (Scharpf, 1987).

New economic thinking is evident in negotiations between government and labor market organizations. Tripartist negotiations on incomes policies characterized the previous periods, but more recent negotiations between the government and labor market partners have concentrated on the unruliness of the labor market, e.g., wildcat strikes, conflict and competition between unions over members, and inflationary wage increases. The aim has been to encourage unions to exercise wage restraint without compensation from tax breaks and special governmental measures. Threats of economic sanctions have replaced the non-zero sum agreements of earlier incomes policies (see Ahlén, 1989).

6. *New Political Actors.* Corporatism as a Swedish style of decisionmaking worked well when there were only a few collective action organizations of political weight. As Swedish society became more complex, new groupings of people began mobilizing support and using their organized voice to put new and different demands on government. The first important new grouping to arise since the period of the Swedish model were white-collar unions. Contemporary new groups are questioning the representativeness of political institutions and older, established collective action organizations. The groups of importance here are women, environmentalists, pensioners, immigrants, and the youth. They articulate new issues as well as the shifting interests of members of older, established collective action organizations. Diversity now characterizes Sweden. Feminism and environmentalism are discussed in the next section.

Organizations for pensioners were created of the Swedish model but are now mobilizing support as never before. They are experiencing what can be called an organizational renaissance. In all there are four national organizations with at least 1,000 members (see appendix 10). They tend to mobilize supporters on the basis of social class. The PRO is the largest organization and has close ties with the labor movement; the Swedish Pensioners' Association (SPF) is non-socialist in orientation; the National Association of State Pensioners (SPRF) is the oldest of the organizations, and the youngest organization, National Association for Pensioner Solidarity (RPG), recruits members from the free church and teetotaling movements. Leaders usually are retired officials from collective action organizations. A merger of the PRO and SPR has been discussed, but their upswing in members had made this idea less pressing.

Roughly one third of all senior citizens (approximately 600,000 people) are organized in these organizations. Growth in membership is explained by the increasing number of pensioners in Sweden, their need for interest organization, and dependence on the welfare state for services and pensions. Previously the organizations focused their attention more on such issues as insurance policies and state subsidy of leisure activities for senior citizens. Now they are articulating their demands in much the same economic way as trade unions. Their main issues are pension levels, taxes, and food and housing costs. Pensioners express their views more vehemently these days and even use political threats, demonstrations, sit-ins, and civil disobedience to get their point across to politicians.

All organizations are interested in better contact with government. Discussions on more stable contacts between government and the PRO began in the 1970s. Pensioners have become such an important group for welfare state policy that the social democratic government decided in 1988 to create an advisory council for deliberations. The non-socialist government has had several meetings with them. These developments signal a closer relationship between government and pensioners' organizations. Their status as a large, politically aware group of voters (approximately 24 percent of the Swedish electorate) makes all political parties take notice of them. Nor can they be ignored by large collective action organizations, which have established contacts with them.

Many young people are developing a completely different relationship with collective action organizations. Rather than investing in established organizations, they opt for such alternatives as New Age, postmaterialist, and new religious movements. The situation is serious for established Christian organizations, which experienced a 40 percent drop in membership among people from seven to twenty-five years of age between 1976 and 1985. This can be compared to a 16 percent drop in membership in teetotaling organizations (Carle and Sjöstrand, 1991, see Blomdahl, 1990). Young people are disillusioned with established collective action organizations, in particularly union organizations, for they do not believe that they can exercise influence within them.

The transition which Sweden is undergoing is severely impacting the once highly revered, traditional channels of interest representation. A study from 1988 showed that Swedes in general and young people in particular felt more affinity with collective action organizations representing postmaterial values– peace and environmental groups and such international help associations as Amnesty International– than social democratic network organizations as the LO and KF whose highest priorities have been economic in nature. Also, there is widespread public indifference to all first wave organizations (Petersson, Westholm and Blomberg, 1989). Opinion surveys underscore these results and point to an all-time low in public trust for Swedish unions. Supporters of New Democracy, a new right political party created in 1990 that was in Parliament from 1991 to 1994, have the least trust in unions (Elliot, 1992). A relatively high number of young, male LO members voted for the party in the 1991 election. Unions are experiencing a crisis of public legitimacy. In 1994 scandals forced the LO and TCO presidents to resign. Organized business has capitalized on many of these events by publishing reports on them. Yet the dire economic situation of Sweden in the 1990s has protected unions from large losses of disenchanted members, who see unions more as insurance companies than social movement organizations.

7. Internationalization. Global developments have changed the parameters of Swedish corporatism. Decisions by government to internationalize the Swedish economy have led to greater dependence on economic developments in other countries. There is growing general awareness that many policy problems are more global than domestic. Governments in many countries see solutions for several policy complexes in international cooperation. Perhaps governments sense that they are 'progressively losing degrees of direct control over the global forces that affect them' (Rosenau, 1988, p. 327). The concept of intermesticity– the integration of international and domestic policy– captures this development nicely. Sweden's international dependence is directly underscored in the decision to seek EU membership.

Many collective action organizations understand that intermesticity lessens the autonomy of Swedish government over politics and, as a result, their own influence in policies. They have begun to direct their activities toward the EU government and participate in organizational networks on an European basis. Similar developments are also found in other countries. Lobbying and opinion formation rather than corporatist exchange are the methods of influence on the European level. Repluralization of Sweden and the development from the strong society towards a more independent Swedish civil society fits well with the emergence of European pluralism and global civil society.

The seven reasons for decorporatization discussed in this subsection form cause-and-effect relationships. Objective changes in reality have influenced subjective perception of reality and vice verse. The underlying reasons for decorporatization are economic changes and public attitudes that demand acceptance of diversity in Swedish society. This means that the basis for organized solidarity has changed dramatically. Objective changes and new political attitudes are producing a new zeitgeist in Sweden.

Postmodern condition and new collective action organizations

The present period is distinguished by major transformation of society, economics, and politics. It is a time of great shifting and churning in politics, economics, and society. There are direct parallels between this period and the first one. The reasons for the current transformation are many, but the most important one is the transition from industrial to post-industrial production. Changes in technology, organization of labor, and capital accumulation are revolutionizing the western world. Industrial society was distinguished by such characteristics as hierarchy, homogeneity, power of the nation-state and trade unions, universalism, centralization, class politics, welfare statism, and government interventionism. These characteristics have also defined the Swedish model (SOU, 1990, no. 44, Micheletti, 1991a).

Industrial society has been called fordist (large-scale, mass production) modernity and is contrasted with post-industrial society or flexible postmodernity which rejects the modernistic conviction in universal, linear, material social progress. This major transition towards fragmentation, differentiation, and multi-directional change are captured in such concepts as 'disorganized capitalism' and 'post-modernity'. The terms 'disembedding' (Giddens, 1990) and 'unsettled lives' (Swidler, 1986) capture the feelings of estrangement. People in many countries are seeking a new orientation and understanding for their existence on earth. New social movements play a central role here. They seek to transform contemporary society and create new identities by writing a new social contract.

Perhaps the three most important new social movements in Sweden are feminism, environmentalism, and religion. They represent the fourth wave of organizational development and concern the issues of destruction and individual integrity. These three movements are postmodern movements because they are concerned with individual diversity, difficulties of communication in society today, and complexity and nuances of interests. Each movement has its own notion of self-actualization, personal growth, and citizen empowerment. Also, their causes are international, and a growing number of them accept the workings of market economics and are concerned with participatory democracy and organizational decentralization.

They are new social movements because satisfaction of their causes would lead to a new value or meaning-system in politics, new organization of society, and new political rationality. They are revitalization and revival movements and challenge the traditional conception of Swedish politics as the struggle between a political left and right over the distribution of wealth and power among social classes. Each movement represents an emerging new political dimension.

Social scientists define the green political dimension in Sweden as a struggle between economic growth, material consumption, and large-scale technology versus self-sustaining economy, ecologically-responsible material consumption, and small-scale living (Vedung, 1992). Voter placement on this dimension is frequently measured by questions on automobile use and environmental pollution, but views on nuclear power and sympathy for the Environmental Green Party have played the most important role in defining this dimension (e.g., Bennulf, 1990a). Most likely this measure underestimates the presence of green thinking in Sweden. The selectively open political opportunity structure has generally hindered the emergence of new political dimensions. For example, the close association of environmental organizations with government as well as the greening of established political parties and collective action organizations has encouraged a different development, which has thwarted the emergence of a distinctive green protest movement that meets a united opposition composed of institutions and organizations of the established political order (Petersson, 1991a, 35, Micheletti, 1991b).

Social scientists have not yet begun to measure the feminist and religious political dimensions in Sweden. Voting patterns of Swedish church-goers are known and the presence of a gender gap have been studied (Bjereld and Gilljam, 1991, Oskarsson, 1990, Gilljam and Holmberg, 1993). Lack of serious study on these subjects may in part explain the mistake made by feminists when they claim that their movement is the only contemporary one in which the personal is discussed politically. This issue is also central for new religious movements. Lack of research here is unfortunate because reactions to the environmental, feminist, and new religious movements are interesting examples of how new politics confronts old politics. The remainder of this section explores the general effects of the movements in Sweden.

As in many countries, new religious movement organizations have played a greater role in the public debate than as agents of political change. There is no balance between the public outcry against them and membership in them. They are criticized for having international and particularly U.S. rather than Swedish roots and using non-Swedish methods to recruit sympathizers and mobilize support. There are striking parallels between reaction to them and the reaction to religious movements from the early periods. Some of the most staunch supporters of the first free churches are among those who label new social movements as political deviants.

There have been important changes in how the Swedish establishment reacts to the new religious organizations emerging in the present period. Attempts to control them are non-violent and less arbitrary, and followers of deviant religions can no longer be exiled, though they are still viewed as religious, social, and political outcasts. Freedom of religion is now constitutionally guaranteed, but public reaction shows that postmodern tolerance of diversity has not yet become a common Swedish characteristic.

What threatens established Swedish society the most is the successes which new religious organizations have had in mobilizing young people at a time when established Swedish churches are stagnating. This implies that new religious movement organizations tend to be more in tune with the current transformation of Sweden. They reflect and reenforce the emerging market-like conditions of modern political life and increasing prominence of religious pluralism or 'spiritual supermarkets' (Beckford, 1985, p. 289). Also, many new religious movement organizations were founded by people of fundamentalist Christian faith who view the established free churches as corruptions of authentic Christianity , a criticism that echoes the view held by the first free churches of the Swedish church. Other new churches are non-Christian and were founded by immigrants in Sweden.

The feminist movement is the least organized of the three new social movements analyzed in this section. Feminist organizations, such as the FBF and a handful of professional associations for women, do exist but most organized feminists now belong to network groups that lack sufficient stability and structure to be classified as true social movement organizations. Examples are: a network for feminist trade unionists which, in a way unique for Sweden, recruits supporters from the LO, TCO as well as the SACO; a network of women in the LO called the Gals' League; Able Women, a foundation to improve the self-confidence of individual women and promote female business enterprises; Thousand Sisters that organizes cultural activities and pressures government for improvements and benefits for divorced women; a short-lived group for women opposed to wage earner funds, and an organization that promotes women's civic representation. The cross-union network folded recently because the TCO incorporated its demands in its goals.

Most feminist consciousness is characterized by either network involvement or individual/private initiatives, which shows that women have a weak power position as an organized interest in Sweden (cf. Hernes, 1987). A sense of urgency for feminist interests is necessary to mobilize feminist consciousness into more organized forms of collective action. There are signs that such a shift in involvement is now taking place in Sweden. The eventuality of EU membership has led many women to question the EU's commitment to gender equality; welfare state services and the rights of working women are of primary concern here. Recent studies have shown that women are more negative to EU membership (Lindahl, 1992). Reform of the Swedish welfare state is even causing women to reassess their political sympathies, as they view strong central government as a crucial factor in their lives (Siim, 1988). Women are more dependent on state services, positive to the public sector, and tend to vote for the political parties whose platforms are more favorable to the welfare state (Oskarsson, 1990, 1992).

Many women feel that they have been let down by political allies, who are either experiencing a loss of public esteem or have shown themselves to be less dedicated to the feminist cause than assumed or hoped by women sympathizers. Erosion of social democratic hegemony is impacting Swedish women. Media coverage and the public debate of the very derogatory comments made by the LO president in late spring of 1992 about the social democratic women's caucus have led many women to conclude that the LO and possibly many other established collective action organizations are not truly dedicated to the cause of gender equality.

Gender equality has been a political priority for twenty years. Organizational units for women's rights have been established on all government levels and in virtually every collective action organization. Government has allocated money for numerous studies on the position of women in society and financed the posts of femocrats and feminist emancipation workers in various political and organizational settings. Their role is to devise schemes to implement gender equality. An association for femocrats and feminist emancipation workers was even established in 1987 and now has about 850 members. These efforts have not fulfilled feminist expectations, and women are now frustrated with the Swedish political establishment. They have participated in collective action organizations and political institutions for decades but are now wondering whether participation can lead to 'women-friendly' politics that redefine the separation between the personal and political sphere. Corporatist collective action organizations are seen as part of the problem for they are oriented toward patriarchal values. Public discussion on this highly controversial issue has, as in the case of new religious movements, revealed a weakness on the part of established political actors to deal adequately with postmodern questions. These developments explain the new mobilization of women into networks. The new social democratic government has promised to give gender equality high priority.

The Swedish environmental movement is the strongest in terms of organizational resources and the most stable of the three movements discussed in this section. It is currently the most politically legitimate. About fifteen national movement organizations exist in Sweden today, though they tend to disagree on ideology, strategy, and tactics and compete over members. A considerable number of sympathizers are members of more than one organization. Incorporation of environmentalism in government has led to tension between a fundis (adamant) wing with an extreme sense of urgency for ecosystem issues, and a realos (pragmatic) wing, which adopts a parliamentary orientation, professional identity, and technical or instrumental version of the environmental message. Environmental consciousness increased dramatically in Sweden in the 1980s. A national political party and numerous local ones now exist. The Environmental Green Party was founded in 1981 and entered Parliament for one term after the first environmental election (1988) in the history of Sweden and again after the 1994 election.

It is a difficult task to summarize analytically how these three new social movements relate to their members, government, political parties and other collective action organizations, and the mass media. The Christian new religious movement stands out as the most deviant: it has received the most public criticism; its members are more isolated from mainstream society, and it has few ties with government. This movement creates difficulties because it challenges government to live up to such liberal democratic values as freedom of religion, speech, and association. It is also the least dependent on government action, though official sanction is needed for its schools. Presently it is the most market-oriented, which makes it a particular target of criticism from adherents of the Swedish model.

The environmental movement is part of the nature conservation government bureaucracy though it challenges and forces the bureaucracy to adopt greener perspectives. This goal explains the concentration on opinion formation and mediazation of events as movement activities. The aim is to heighten the environmental consciousness of the Swedish people. There is a close affinity of thought with the Environmental Green Party, but many environmental movement organizations have chosen to work independently of it. Greenpeace and even the largest environmental organizations, the SNF, stresses market-like over government regulation as a way of solving environmental problems.

Greenpeace is seen as the most deviant of Swedish environmental movement organizations and has been criticized for not conforming to Swedish norms for collective action organizations. It would seem as though political deviance here leads to greater legitimacy, for opinion surveys have show nthat it rates highest on public respect for environmental movement organizations (Bennulf, 1990b).

The feminist movement has been the most conventional of the three movements, though this situation is now undergoing change. Methods for opinion formation are only in a rudimentary stage of development, and market-orientation is not even a topic of discussion. It is more dependent on government for goal fulfillment than the other two new collective action organizations. It differs from the environmental movement in that it has only recently begun to distance itself from the political establishment and to react to recent changes in the political opportunity structure. The crucial issue is whether it is possible for feminist interests to be satisfied through traditional channels or new strategies for collective action.

Do established collective action organizations have a future?

Postmodernization is impacting Swedish collective action organizations and civil society in major ways. New ideas and forces from postindustrial and postmodern experiences confront old conceptions and actors whose origin is found in industrial society. Not all established collective action organizations have been able to adapt themselves to these changes. The adjective 'new' has even become the ideological property of neo-liberal forces and the counterculture movement of the 1990s. The labor movement and many older collective action organizations have not been able to formulate a viable alternative to this challenge. Political tensions between old and new forces affect collective action organizations internally. There are struggles within almost every organization between traditionalists or defenders of older conceptions and procedures and people wanting renewal and change.

Probably the most important example of the impact of general change is the decline of trade unionism. Today unions are less important politically, and organized business is successful in influencing politics. Other political actors are taking over the role once played by unions. For instance, the SAF's reform proposals on deregulation and decorporatization of Swedish public policy, privatization of state companies and welfare services, and decentralization of collective bargaining are in tune with and actively framing the dominant interpretation of current political realities. A new balance of political bargaining power is emerging. One significant aspect of this balance is new thinking about capitalism. Today market relations are acclaimed by a growing segment of society as the main instrument for reform. They see market-like relations as facilitating change, offering alternatives, empowering citizens, and leading to pluralism as well as renewal of established collective action organizations. This contrasts greatly with the ideology of the Swedish model in which regulation and the planned economy were seen as proper government tasks.

This line of thought is captured in conceptions of citizens and members who place demands on the services offered by government and collective action organizations. Members are customers who engage in comparative shopping, which implies that there must be more than one organization for the same kind of collective action so that customer-members can compare their merits and decide whether they wish to accept the ideology and services of a particular organization. Adherents of marketization argue that market-like relations empower organizational members because they can express dissatisfaction by leaving the organization (exit), complaining (voice), or accepting the existing situation (loyalty) or combine the three alternatives.

Many Swedish collective action organizations in the western world are now going through a marketing phase of organizational development. This does not, however, mean that they are not skeptical about its virtues. Certain of them are studying how they can apply market-like logic to renew their organizational structures and democracy. The Union of Clerical and Technical Employees in Industry (SIF) is of particular interest here because it is the first union organization to attempt to integrate business strategy using 'methods to affect industrial relations outside the bargaining area' (Reshef and Murray, 1988, p. 92). It began to market the worth of union membership to check a 20 percent loss of members in the Stockholm region and has constructed a system of union membership which allows members to put together an individualized union coverage package. The idea is that the most inexpensive membership package covers the essentials of Swedish unionism– collective bargaining, economic advice, and certain help with career opportunities– while the most expensive one entails legal aid, individualized help in collective bargaining, welfare and insurance assistance, and other services that focus on the individuals's career opportunities and work situation. Only a few Swedish organizations completely base their relationship to members on market-like logic. Greenpeace offers potential 'checkbook members' an ideology for sale that they may buy for a longer or shorter period of time. It is more common for organizations to offer fringe benefits– e.g., special offers on insurance and various products– to entice new and keep old members (for examples see Micheletti, 1988a, Ahlén, 1990). Others are discussing how to make membership more contractual.

A second aspect of the new societal balance is privatization. Few if any interest organizations or social movements have not created stock market companies for member services. Some organizations have virtually become private enterprises. Today it is more correct to refer to the consumer and farmers' cooperative movements as capitalist cooperatives (Micheletti, 1987, Holm, 1989). The reason is Swedish laws for economic associations, which make it difficult for cooperatives to subsist in the harsher economic climate of this period. A few enthusiasts of privatization have even proposed that several of the most revered and cherished activities of established collective action organizations be transformed to movement stock market companies. A related problem here is the increasing number of self-employed individuals who retain their union memberships. Structural transition of production has blurred the distinction between employers and employees. One wonders whether it will, in the near future, also blue the distinction between unions and employers' associations.

A third aspect is decentralization. As in the case of marketization and privatization, trends in government have preceded developments within collective action organizations. Decentralization became very important for labor market organizations when the SAF decided that it would no longer negotiate centrally with the LO and central government began to transfer welfare services to regional and local governments (see Ahlén, 1989). This has impacted unions, which are now making an effort to localize their activities. Collective action organizations are focusing attention on decentralizing decisionmaking authority to regional and local units. The aim is adaptation to the new political opportunity structure that decentralization has created and improved contact with members.

A fourth part of the new balance is individualization. Unions concentrate on 'the individual within the collectivity' and 'the individual in focus'. Both slogans signify a conception of solidarity that rejects standard or universal solutions and promotes flexibility and individual diversity. This development has been called new or consumerist individualism and reflects postmodern, market-like tendencies in society. The solidarity of the Swedish model was instrumental, elitist, and more similar in fact to loyalty to political and organizational leaders than true solidarity as defined as communality of interest (see Micheletti, 1990b). Self-denial or loyalty to the organizational cause characterized the concept of solidarity in the period of the strong society and Swedish model. The new solidarity developing now in postmodern Sweden is more inward and self-fulfilling. Young people are more new individualist than older ones, a development that has led to worry within several collective action organizations which tend to view this attitude as anti-system. There is concern for depoliticization (loss of interest in established politics) and apoliticization (active distrust in established politics) among younger citizens (cf. Mushaben, 1983). Organizations that have not been able or willing to adapt themselves to new individualism are losing members and finding it more difficult to recruit younger people. First and second wave organizations are the best example here.

Many established organizations find that a growing segment of the Swedish population is opting for private/individual rather collective action to satisfy their needs. There is concern for private teetotaling, private sporting or new sport activities as jogging and weight lifting that do not require organizational membership, and private religious worship or nonchurch belief in a god. This reflects a shifting involvement and possibly a new way of viewing the separation of public and private spheres of life.

New individualism poses important challenges for established collective action organizations, which must be willing and able to answer questions about what passive membership and active involvement in the organization give the individual and how membership promotes the organizational cause directly. Labor market and other organizations devote considerable resources these days to conferences to understand the impact of new individualism as well as the needs and desires of young people on their traditions and practices. At the same time, Swedes express a lower level of confidence in union leaders. Research on the white-collar unions has shown that new union issues cannot be satisfactorily solved at central levels and need higher level of local commitment from members. The problem is that this kind of commitment has defied the traditional conception of interest representation (Boglind, 1989).

Similar problems in the relationship between members and organizational leaders and officials, as well as with member commitment and organization structure, have been experienced by other collective action organizations. Konsum Stockholm, a member of the KF, considered these problems so severe that it appointed a committee to investigate its member democracy and propose solutions to its problems. The committee reflected on the conception of democracy characteristic of such established organizations as the KF. It believed that too much emphasis has been put on formal elements of democracy, i.e., rules of order, which it believed provided a 'terrible setting for discussion and communication and was not method for influence which many members could use successfully' (Konsum, 1988, p. 16, my translation). The committee encouraged Konsum Stockholm to consider a new approach which it called 'membership opportunities' and defined as including three key concepts: interest, participation, and influence.

Few established collective action organizations have changed their attitude and behavior. Some organizations, as the ABF, ask whether they can be young again, and certain LO functionaries acknowledge that young people cannot be reached by inviting them for coffee and donuts in the movement's local community center (Närling, 1993a). Organizations that can successfully appeal to the sense of solidarity and community felt by people socialized by new individualism will be able to have a significant impact in the future. New religious movement organizations satisfy the longing for a sense of community for a small segment of Swedish society, and humanitarian and new social movements, local actions, and civil disobedience appeal to people who want to feel that their individual participation matters.

A fifth aspect of the new balance is internationalization; its implications for collective action organizations were discussed earlier. The five aspects discussed here signal major transition of the role that collective action organizations play in Swedish politics. The problem is that many collective action organizations have difficulty adapting themselves to them. A main reason here is inflexible thinking and structures, a characterization of the life cycle phases of stagnation. The question is whether Sweden's collective action organizations can renew themselves. Most if not all organizations have or are undergoing some kind of structural and ideological renewal. Structural renewal involves rationalization, decentralization, and mergers. Ideological renewal concerns activating old and recruiting new members and formulating goals that reflect the zeitgeist. The three most important groups of new members are those individuals which have been neglected by collective action organizations: women, young people, and immigrants.

Organizations discuss both types of renewal, but it is quite obvious that structural renewal is given priority. Ideological revitalization is a sensitive topic, and traditionalists within organizations are reluctant to allow serious discussion of it. A major problem here is lack of good, enthusiastic leadership and increased heterogeneity within the membership ranks, two developments which make it very difficult to agree on the future goals of the organization. Ideological renewal is complicated by corporatist forms of interaction, close ties with other collective action organizations and political parties, and tendencies toward oligarchical leadership. The blue-collar union movement is the best example here.

Many of these problems were present in earlier periods but have become more serious as the economic situations within organizations worsens. Member organizations and local units are insisting that they be allowed to keep a greater share of member fees for their own activities, and these demands are leading to decentralization. This even applies to the most encompassing and resourceful collective action organizations including the SAF, SACO, TCO, and LO. The situation is quite severe: members oppose union fees increases; parliament voted to cancel the tax deduction for union membership which was introduced in 1989, and the general economic state of Sweden is not good. Other collective action organizations rely on government patronage to cover budget deficits, but there are serious problems involved here as government is reconsidering its financial commitment to collective action organizations. Reliance on non-government external sponsors is an alternative. The dilemma for collective action organizations is whether they want to be member-based and rely on member involvement for continued existence or dependent on an artificial life-saving system provided by government and other external sources.

Peak organizations are studying how costs can be reduced; rationalization and mergers are common proposals. There have been numerous and extreme mergers of unions in earlier periods of Swedish civil society history that aimed at centralizing power to the peak union organizations and constructing viable industrial unions. Centralization and industrial unionism characterized the Swedish model. Contemporary mergers are motivated by economic hardship, conflict among organizations over members, and internationalization. Government privatization is changing the boundaries between the private and public sectors; new technology is blurring the distinction between blue- and white-collar workers, and there is a need for unity as Sweden takes a more active part of the EU.

Unions that represent different employment sectors within the TCO are discussing either mergers or more intimate contact; private sector TCO unions are contemplating merging; LO unions from declining sections of Swedish industry have merged, and there is serious talk about creating a large peak union that can represent all employees in negotiations with the EU by merging the LO and TCO. TCO unions were the first to consider and implement new mergers.

Even employer associations are merging and reevaluating their structures. It is possible that the SAF and IF may organize certain services jointly or merge completely in the near future. Government privatization has led the Employers' Alliance, first created in 1970 to represent public sector employers against unions in collective bargaining sessions and later broadened to include the sports, cooperative, and environmental movement organizations and privatized state companies, to encourage its members to join the SAF.

Mergers, structural rationalization, and developments of organizational networks are discussed within other collective action organizations as well. Free church movement organizations have over the years considered pooling their resources. Joint activities have been rather common, but ideological differences have tended to hinder closer and more integrated cooperation. Significant change in attitude occurred in the 1980s, due to frequent cooperative efforts on the grassroot level. Economic problems, declining memberships, demographic changes, belief in ecumenism particularly among young members, and the need to reflect the zeitgeist are the reasons for increased support for mergers. A new sense of Christian community is emerging, though identification with individual free churches has declined particularly among younger members. A plan exists for merging the Covenant, Methodist, and Baptist churches, and it is not impossible that the Free Baptist, Pentecostal, Word of Life, and two smaller free churches will cooperate more closely in the future.

Mergers and creation of organizational networks are provoked by economic necessity, but they also signify a search for new collective identity formation, as shown in the case of the Swedish free church movement. Today many collective action organizations want to extend their goal agendas into less traditional fields. This may lead to structural as well as ideological renewal. New contact triggers new expectations among members and provides cooperative learning experiences. Development along these lines reflects the desectorization of Swedish politics and society. Examples of new networks include cooperation between: the peace and labor movement; unions and immigrant organizations; teetotaling movement, unions, and other collective action organizations which specialize in foreign aid projects; the HSB, SNF, and women's, peace, and environmental movements; Christian free churches and environmental organizations, and, finally, cross-union networks for women. Attempts are made to improve contact between new and old social movements. It should be noted that these new networks are reminiscent of the ones from the first and second periods.

More spontaneity within organizations is sought by concerned members and leaders. Bureaucratization allows standard operating procedures to dominate organizational work, and corporatism encouraged and forced organizations to accept more public responsibility for their actions than has been beneficial for their well-being. It encroached on civil society by placing too many demands on collective action organizations. The result is alienation, misgivings, sluggishness, inflexibility, and oligarchy. Organizational activists have become cynics, who calculate what activism can give them personally. Several leaders now proclaim that their organizations must return to their roots as protest movements (Micheletti, 1988a).

Contracting out was proposed in the 1970s and is intensely debated. The issues are the size of the welfare state and vitality of Swedish social movements. Many municipalities began to transfer certain services to social movements in the 1970s. Examples are leisure time activities for young people and maintenance of ballparks. Both local government and movement organizations have been satisfied with this development. Service transfer helps government cut costs, and financial assistance for administrating services helps fill organizational treasuries. Also, contracting out is seen as a step towards more direct local democracy, and it gives collective action organizations an important civic task. In the 1990s, there is even talk of funding voluntary organizations to run refugee camps. Economic necessity and democratic thinking seem to be going hand-in-hand for local government and social movement organizations.

Few political actors question whether contracting out is beneficial for social movements *per se*. Yet as discussed in the previous chapter, experience with young people shows that acceptance of responsibility for a social service can tax collective action organizations greatly. Does it lead to government control of organizations? The author of a government study on this matter concluded that this indeed took place: 'Local governments have on occasion persuaded social movement organizations to take over services that have transformed the operation of the organization. In this sense, the municipality redefines the activities of the organization...Municipal blackmail is a strong term, but this kind of behavior is more similar to blackmail than contracting out' (Ström, 1985, p. 50, my translation). Given this evaluation, it is thought-provoking that the government became more interested in contracting out in the 1980s and even appointed a special commission to investigate the matter.

Public debate on the role of collective action organizations in the welfare state in the 1990s is less intense and concerns the non-profit sector. Proponents of voluntarism want to provide opportunities for citizen participation in and responsibility for social services. The aim is citizen empowerment and strengthening of pluralist and democratic forces in society. Swedish proponents see voluntarism as a way to revive stagnating collective action organizations and help solve new social problems.

That a non-profit or volunteering sector has reemerged in Sweden is shown in the increasing number of collective action organizations. The Swedish National Tax Authority recorded 30,936 voluntary and 76,064 economic associations in the 1980s, with 12,015 newly registered voluntary and 38,479 economic associations in 1980s. This reflects an increase in 39 percent for voluntary organizations and a 51 percent increase in the number of cooperative or economic associations, many of the latter are housing cooperatives. Few collective action organizations exist outside this register.

New groups have emerged to deal with social problems with which the welfare state cannot cope; many of them have the term 'ad hoc' in their title. It is too early to assess whether they will become organizationally stable and develop into social movement organizations. The greater portion of them deal with specific problems. Several are anti-violence groups which struggle for human and even animal integrity. Some groups are NIMBIES and want to keep destruction from their neighborhoods. Destruction may be defined as pornography, air pollution, or desire to prevent construction of an airport in their neighborhood. The struggle against human solitude takes organizational form in groups for divorcees, and there are even groups which promote sociability. Many ad hoc groups have started for specific illnesses. Several ad hoc groups are concerned about national and personal economic status. New groups are also emerging as countervailing powers. There are groups for improving public transportation and those that fight for private automobiles as well as those for and against tax cuts. The abortion issue is also leading to collective action for people who are opposed to current Swedish abortion policy.

The internationalization of Sweden has also led to the creation of numerous groups. Campaign organizations for and against EU membership mobilized support for the national referendum held on November 13, 1994. Established collective action organizations are creating networks with system organizations in other European countries.

Some of the established collective action organizations involved with humanitarian aid are finding that they have renewed public legitimacy (see appendix 9 for membership statistics). Membership in the Swedish Red Cross is, for instance, growing. Many young people are new supporters. Private enterprise and the public is showing new interest by contributing financially. Voluntary associations have even established a foundation to collect contributions and recruit sponsors. The non-socialist government contributed financially to the creation of a center for voluntary associations. Save the Children Foundation, the Red Cross, free church and teetotaling organizations, and others are a focus of mass media attention. The organizations prefer not to be labeled as charities and point to experience from other countries where professional volunteering sectors are beneficial for society and democracy. Concern about the costs and size of the Swedish welfare state is an important reason for the boom of voluntarism in Sweden. New volunteering organizations have also been created. Also, the organizations are moving into new areas not sufficiently covered in government welfare policy. Most political parties see this development as a key to community-building, civil republicanism, and an act of solidarity. It is, therefore, possible to speak of a fifth wave of organizational development in Sweden.

The examples show that establish collective action organizations can adapt to the zeitgeist. It is possible for established collective action movements to renew themselves structurally and ideologically. The inability of the welfare state to cater to the needs of old and new groups of citizens has given collective action organizations a new social call and pertinence, which is reflected in public commitment to voluntarism. A new sense of public responsibility-taking is developing in Sweden. It does not concern material welfare and compromises with government as in the periods of the strong society and Swedish model. Rather, it involves helping individuals to live a better life, and it gives members personal satisfaction from contributing to a good cause.

There are clear parallels with the earliest periods when collective action organizations mobilized support to help the materially-needy. Now they mobilize support to help Swedes with problems caused by material welfare and non-Swedes who are victims of destruction. The postmodern condition is shifting the involvement of Swedes towards the volunteering sector and leading to a resurgence of Swedish civil society. The trend is definitely towards empowered citizens to be less reliant on government and to take responsibility for shaping their individual and common future.

8 Over a century of collective action organizations

Social movements and interest organizations have played an important role in Sweden since they emerged in the 1800s. The previous chapters show that individual organizations have played varying roles, all depending on their relationship with the political order and internal circumstances facing them. The emergence of new organizations in several of the periods compelled older organizations to clarify their stand on the political order and react to newcomers that challenged them. The relationship between collective action organizations and their members has also changed over time. The key here is how accumulation of different resources have impacted collective action organizations and furthered their development towards institutionalization and maturity. Member commitment to several of the organizations of the first wave peaked in the second period. Support for them has been born by other patrons. A very important one has been Swedish government.

The political opportunity structure of Sweden has undergone transformation in many periods. Each period has witnessed change in the working relation between government and organizations. It is worthy to note that the two significant ages of transition, which are represented by the first and second periods (1850s to 1920s) and fifth and sixth periods (1960s to 1990s), brought about crucial changes in the relationship between government and civil society. The new confronted the old in the late 1800s and early decades of the 1900s and changed Sweden significantly. A similar process is occurring in the late decades of the 1900s. The key developments at the time Sweden approached the twentieth century are summarized in the terms democratization, internationalization, individualization, and industrialization (Lundkvist, 1977). The turn of the twenty-first century is characterized by similar terms, individualization, internationalization, decentralization, marketization, and post-industrialization (SOU, 1990, no. 44). This chapter takes a thematic look at the continuities and discontinuities in Swedish civil society history.

Political zeitgeists, renewal of civil society, and organizational maintenance

Emphasis in this book has been put on understanding Swedish civil society history in terms of periods of collective action, organizational waves, and organizational life cycles. Can conclusions be drawn about their interconnections? Are developments in civil society, as summarized in the six periods, in any way related to organizational waves and life cycles? Do times of social, economic, and political transition require special kinds of collective action? What role do political zeitgeists play in influencing the way people perceive their private and public involvements in the affairs of their country? Times of great transition are periods of new beginnings and renewal. New demands emerge, and new identities are formed around them. These identities concern individual, social, economic, and political citizenship. Segments of the population may find that older identities and forms of collective action are unable to respond to their current and urgent needs, as the sense of justice, democracy, and equality of older organizations is delimited by their goals, organizational structures, and relationship to the political order. Mature organizations which cannot adapt themselves begin to stagnate, as their own concerns are given lower priority by both members and the general public. Newer organizations emerge and gradually take their place. This is the general way in which political zeitgeists and organizational waves and life cycles interrelate.

An example will explain this interrelationship. Swedish blue-collar unions were created in the industrial revolution and define their demands in terms of class conflict, which has deemphasized the importance of gender and ethnic identities as well as postmaterial concerns. The future of unions will be determined by their ability to incorporate the new demands of diversity and recognition in their goals. Yet organizational cultures and structures as well as democratic procedures and practices work against such renewal. Swedish unions and many other collective action organizations are trapped in the zeitgeist of the Swedish model and find it difficult to disembed themselves from its assumptions and behavioral patterns.

Regardless of whether changes are based in structural developments or subjective perceptions, they shape the political culture of a country. Each period has its own conception of homogeneity and struggles of recognition. The issues in the first and second periods were those of the first and second waves of organizational development, i.e., political and religious freedoms and recognition of social class and gender as political issues. The collective actors of importance here are women's, free church, temperance/teetotaling, peace movements, and unions. They played a central role in the formation of modern civil society. Main issues of recognition in the second, third, and fourth periods concerned granting economic rights to disadvantaged groups in society and empowering them as citizens. The collective action organizations that fought for these rights became institutionalized and later mature. They created the Swedish model of politics, which incorporated economic citizenship as its basic tenets. Economic redistribution became a fundament of Swedish politics, and non-class-based collective action organizations found that their issues were given lower priority in politics.

Another development characterizes the fifth and six periods. The issues of recognition differ tremendously from those of the Swedish model. Questions were and are raised about the political wisdom of solving complex social, economic, and political problems with standard solutions and class analysis. The elite democratic political culture of the Swedish model is challenged by groups demanding a say in politics. They want the Swedish people to sign a new social contract that includes their demands and political interpretations. Among the organizations involved here have been the peace, environmental, student, and women's movements, NIMBIES, and voluntary associations.

As time went on, many of the social movements that were agents of democratic change in the first and second periods became more similar to interest organizations. They changed from being more like true social movements– with members dedicated to the urgency of the cause and willing to confront the political order to enlarge the boundary of the political through protests– and became more similar to interest organizations, due to their concern about political legitimacy and organizational maintenance. They fought against new collective action organizations that sought to change politics in ways which challenged their basic structures and traditional goals. The social movements which emerged as Sweden's modern civil society took form are now more similar to self-interest collective action organizations. Many of them are more reliant on government grants than member support for survival. Some of them became part of the corporatist apparatus and lost their character as independent organizations of civil society.

Figure 8.1 summarizes the findings of this book on the history of Swedish collective action and formation, collapse, and resurgence of Swedish civil society. Emphasis is put on the two ages of transition, as represented by the first and second as well as fifth and sixth periods. The figure shows how collective action organizations have developed along the life cycle model, how the political order has reacted to new political ideas and actors over time, and how the periods discussed in this book and organizational waves are related with each other. It should be noted that organizational waves tend to overlap in that there are not very distinct ends and beginnings of sequential waves. The same is true for life cycle stages and periods of civil society history.

Figure 8.1 shows that even resourceful and mature collective action organizations begin to stagnate once the political order is challenged successfully by a new zeitgeist. This happened in pre-modern Sweden in the first period and to the Swedish model in the fifth period. Decline of the Swedish model as an effective decisionmaking tool led to stagnation for such mature and resourceful collective action organizations as unions and employers' associations. Inefficiencies in and problems created by the Swedish model have led to limits on government expansion as well as transfer of responsibilities to the volunteering sector, which includes such mature and possibly stagnating organizations as the Swedish Red Cross and small, non-institutionalized organizations concerned with the welfare of diverse social groups. The figure also raises an important question about the future of Swedish collective action after the period of the resurgence of civil society. Will Sweden follow a similar path of development as in its past? Will a new strong society be created, and can a Swedish model develop out of the new and different circumstances that have been created by the great transition to postmodernism? These issues are debated today.

A new Swedish model must take into consideration the kinds of demands, interests and collective action organizations that are now making themselves heard in politics. Contemporary issues of diversity and recognition must be given a role. The old pillars of Swedish politics– organized workers, farmers, and employers– must play a less dominant role if the new political actors are to be given an influential role in deciding the future.

Also, it may be asked whether cycles of collective action, as formulated in figure 8.1, reoccur. Do great transformations of politics, society, and economics and significant organizational waves occur in hundred year cycles? A longer historical perspective is necessary to answer this question properly. Neither is it possible to answer definitely whether great ages of transformation are followed by periods in which the changes of the ages are consolidated, institutionalized, and modified to suit the traditions of national political culture. Sweden experienced a period of consolidation after the first and second periods, and it is quite possible that a similar process will occur after the sixth period. It would seem that organizational waves follow a similar cycle: significant waves are followed by more minor ones, which accommodate protest demands to the political culture and opportunity structure. Even organizational life cycles seem to follow this development, as genesis is followed by stages of structurization and management of ideologies and causes.

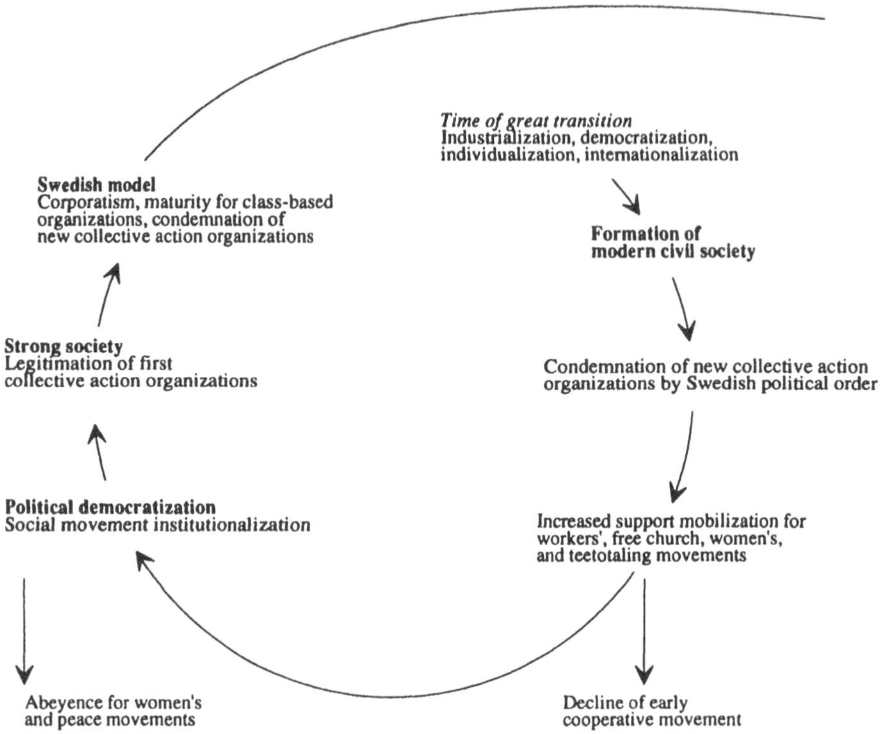

Figure 8.1 Cycles of Swedish civil society history

Powerful allies, opinion formation, and role of leadership

All collective action organizations need support to be effective. Sympathizers are individuals, political parties, other collective action organizations, government, private business, and the mass media. This book has shown that all categories of supporters are important, though not of equal significance for all organizations and in all periods. Direct contacts with the public are more important when a high public profile is necessary. Opinion formation is particularly crucial for support mobilization in these times. It is used to court potential powerful allies, convince traditional allies that their support is still needed, and influence the public to sympathize with the organization and its cause. In stable economic, political, and social periods, collective action organizations can rely more on regular, less public contacts. Contact with the

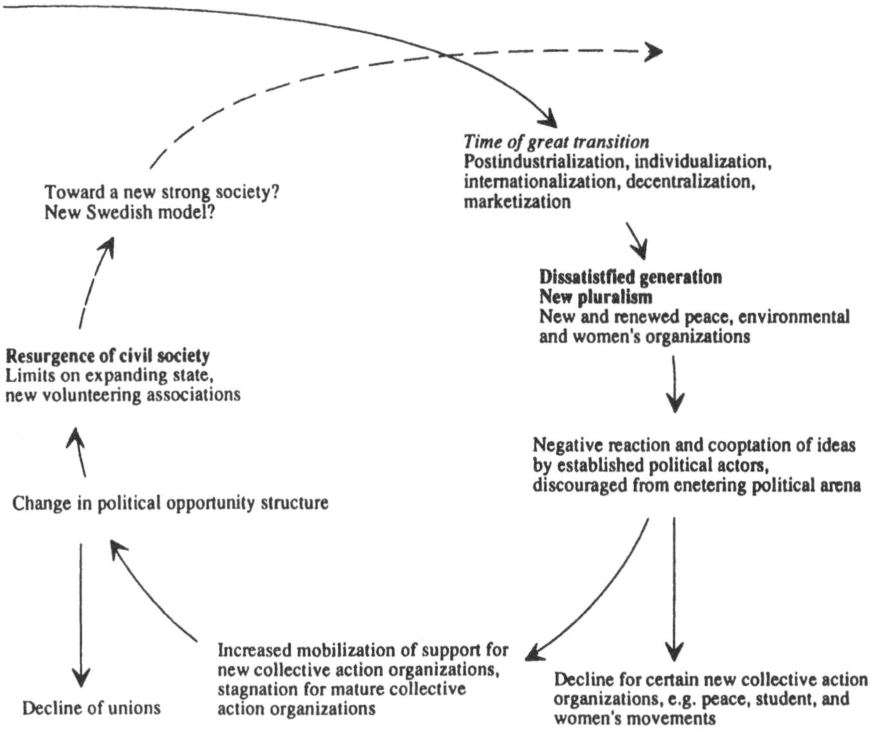

mass media and mass media coverage of collective action organizations is nothing new. This explains efforts by the first social movements to create events.

Mass communications gradually replaced a central role played by certain organizations in public debates. Teetotaling societies and adult study associations typically functioned as discussion or debate clubs for their supporters. In many ways, the mass media has taken over this role with its talk shows, documentaries, and news programs. Also, journalistic professionalism now signifies that the mass media is no longer just an arena or channel for broadcasting news as defined by political actors. Organizations can no longer rely on journalists to relay their press releases to the public. Journalists assess such information after their own professional standards and create news. These developments have greatly impacted collective action organizations in the fifth and sixth periods and led them to invest more resources into direct opinion formation. All organizations are now aware of the importance of opinion formation for support mobilization; many of them have professionalized their press, information, and PR units and have developed or are developing media strategies to exercise political influence (Närling, 1993b, Micheletti, 1985a).

Leaders play a crucial role in adapting their organizations to new general developments. They have always been important for mobilizing supporters. Their general task is to create bonds with allies, and they are ultimately responsible for the positive and negative public notice that their organizations receive. Leaders can be likened to managers and problem shooters, who deal with intricate, concrete problems involving internal organizational affairs, and foreign diplomats, who mediate between the organization and its external environment. They represent and symbolize their organizations and attract attention to and explain organizational policy and goals to supporters and non-supporters.

The life cycle model of organizational development conceptualizes the interrelations between leadership and such other important organizational characteristics as member commitment and political legitimacy. Leaders that entice general support for the organization are less important when the organization is involved in a trustful partnership with powerful allies and when internal disagreement is low. Their ability to communicate successfully with and through the mass media becomes crucial once this situation changes.

Today the task of leadership is extremely difficult as Sweden undergoes the process of decorporatization. Many mature collective action organizations grew large and became stable and politically legitimate in the corporatist periods. It is very apparent that corporatist power sharing has had negative effects on them. Among other things, it led to a widening of the gap between members and leaders. Trade unions have suffered the most: they began to discipline their members to accept the dictates of leaders; leaders became mediators between government and member interests as well as independent forces within the organizations themselves. The consequence that we see today is growing mistrust within many collective action organizations. This kind of apprehension explains difficulties with renewal and oligarchical tendencies within organizations themselves.

The internal and external developments in the final two periods have altered the needs for leadership within collective action organizations. Leadership styles well adapted for the corporatist exchange are ill-suited for renewal. Today many organizations are attempting to deal with these problems through leadership renewal and opinion formation. Several organizations offer leadership courses, and some of the largest organizations have created leadership institutes. A report from the LO stated blankly: 'If we cannot cope with the role of leadership, we are not going to be able to deal with our union tasks in the future' (LO, 1989, p. 23, my translation).

Individuals who can plan strategically and tactically, choose the right alternatives, formulate visions for the future, adapt the organization to the current zeitgeist, deal successfully with the mass media, and create networks with allies while remaining accessible to and representative of members and other supporters are seen as having the necessary leadership qualities for the 1990s. Individual leadership is now reclaimed as important. The hope is that new leadership will reflect the diversity of Swedish society, need for participatory democracy, and prepare collective action organizations for the great transformation of Sweden brought on by individualization, marketization, decentralization, internationalization, and postindustrialization.

The expression 'team leader' captures nicely this trend. The role model here is the leader of a sports' team who works closely with members, activates them, gives them criticism and praise directly, incorporates their ideas and complaints in the organizational dialogue, and solves problems immediately and non-bureaucratically (Närling, 1989). Team leaders combine leadership qualities from different phases in the organizational life cycle. They resemble pioneers and agitators from the genesis phase, for they have ideas, courage, and fantasy. But they also know that large organizations also must be run properly, and political legitimacy must be maintained. Good leadership is crucial for organizational opinion formation and network-creation. Leaders are the people engaged in organizing the dialogue internally and externally. Their task is communication; the cooperation of members and other organizational supporters is essential for its successful completion. This means that contemporary leaders are required to combine charisma, statesmanlike behavior, and administrative talents. These leadership traits are characteristic of the genesis, institutionalization, and maturity phases of the organizational life cycle.

Inability or reluctance to evaluate themselves critically has meant that many collective action organizations have refused to face their problems. Examples here are abundant. Investigations of internal democracy within the LO and TCO conducted in the 1970s and 1980s did not deal with the internal democratic problems confronting mature peak labor market organizations. Even journalists employed by organizational newspapers have experienced that critical, constructive viewpoints are coldly received and editors in chief have been dismissed. Leaders have also criticized articles and actions of investigative journalists working for the organizational newspaper.

The internal and external consequences of lack of organizational openness to internal criticism and disagreements between traditionalists and proponents of renewal are becoming very apparent. An author with ties with the ABF expressed his frustration with the defensive organizational maintenance of the labor movements in the following way:

> Haven't educational efforts been geared too much for elected leaders and organizational employees and thereby contributing to the crea-

> tion of an elite within the labor movement, which often has insufficient contact with other members?
>
> Haven't we too often neglected spontaneous expressions of opinion and occasionally dealt with them with rather brusque measures? Hasn't the rather paltry method that the labor movement uses to encourage dialogue and member influence, counselling the members, been misused and mistreated? Isn't it time for us to realize that we cannot keep a social movement alive with only a halfhearted election campaign every three years? (I. Johansson, 1992, p. 10, my translation).

The SAF began in the late 1960s to reflect upon its situation and went through a renewal process relatively early and successfully. The impetus for change was the Swedish political opportunity structure. The process led to a new collective understanding of its role as an employers' association. Among the changes are an offensive strategy for opinion formation, a higher political profile, reassessment of the corporatist exchange, and dismantling of its collective bargaining section. These efforts have been discussed in earlier chapters.

The LRF started the process in the late 1980s and has been rather successful in mobilizing its heterogeneous membership to pull in the same direction, initiating internal and external opinion formation campaigns that encourage dynamic organizational dialogue, reassessing its relationship with the political parties in parliament, and recruiting new leaders. Its impetus for change was political as well as economic, as the economic situation of farmers was effected by proposals for agricultural deregulation. Change in government in 1991 has lessened the sense of urgency here.

The peak labor unions– TCO and particularly LO– have lagged behind, but there is sign of change. In particular the TCO has reassessed its general organizational strategy, and it is a stronger organization today than in the recent past. The LO is, in many ways, worst off, due to its relationship with the Social Democratic Party, but emphasis is now put on renewing organizational structures and adapting organizational goals to the new zeitgeist. The impetus for renewal is basically economic. At present it cannot be determined whether the efforts concern offensive or defensive organizational maintenance. Traditional thinking and general organizational rigidity may hinder renewal.

A key effort in the renewal process is investment of resources in development of mass media and opinion formation strategies. Yet if opinion formation strategies are to succeed, they must involve the entire organization and not just transform leaders into media stars. Also, the tendency throughout civil society history has been for younger organizations to view new tactics positively, while supporters of older ones feel that modern techniques threaten traditional organizational goals. In general organizational supporters have seen the issues of strategy and goals as interrelated, and the issue of contention is whether new media can convey the organizational message without distorting it.

Opinion formation tactics have become more numerous, and collective action organizations invest more effort in their planning. Among the techniques are video films, rock music, film and television commercials, public advertisement campaigns, created events, newspaper distribution, and such commercial marketing efforts as advertisement of the organizational logotype as brand names on t-shirts, plastic and cloth bags, socks, golf clothes, etc. Such politically legitimate and encompassing organizations as the SAF and LO, more controversial and less legitimate organizations (e.g., Greenpeace) use these kinds of tactics as do smaller less well-known ones as Dads and Moms in Stockholm City. Also, many organizations– including the Pentecostal Movement and several other free churches, the Swedish Federation for Gay and Lesbian Rights (RFSL), SAF, and LO– have their own local radio programs. Organizations are also discussing cable television. Stagnation has led many collective action organizations to use opinion formation to target new groups, in particular women and young men.

Collective action organizations have over the years created internal information units and later enhanced their status. The difference is that opinion formation campaigns are now permanent features of their existence: they define organizational action. Influential collective action organizations are either involved in opinion formation campaigns or busy planning them. Flexibility in planning is the key to success here. It is quite probable that decline of Sweden's first social movements is directly related to their inability to plan successful opinion formation campaigns. For instance, the IOGT-NTO has traditionally taken five years to plan an opinion formation campaign and has not put effort into rallying supporters for the organizational cause in the interim period. The consequence is that it fails to take up issues of immediate importance and capitalize on them in the public debate and thereby misses the opportunity to show its supporters and the general public that its cause is urgent.

Inability to relate the organizational message to current concerns and find skilled people to plan successful opinion formation campaigns characterize other traditional collective organizations as well, for instance the ABF and free church movement. Other organizations– e.g., the RFSL and Swedish Association for Aid to Drug Abusers (RFHL)– lack sufficient financial resources to implement PR ideas. The FBF is of particular interest because its problems with creating public notice stem not only from lack of resources but from internal dissention and problems with leadership. This example shows the interrelatedness of leadership and opinion formation (Närling, 1993b).

There are even organizations which have not experienced problems caused by changes in Swedish political opportunity structure. Amnesty International has not altered its tactics, for it is not dependent on Swedish government for goal attainment. The HSB, Study Promotion Association, and two associations of organized business, the IF and Swedish Federation of Small Enterprise, consider the contacts which they presently have with political parties and other allies to be sufficient for them to exercise political influence and mobilize necessary support. They view their traditional means for distributing information as satisfactory and do not sense a need to mobilize support among other than traditional support groups. This was also the position of the HSB until the economic recession of the 1990s forced it to reevaluate its position.

Resources for opinion formation are, without a doubt, the most important organizational resource for goal attainment in the 1990s. This is not only explained by the role of government and therefore voters for organizational goal attainment. Rallying support for collective action organizations also plays a significant role in times of great transition, when the boundaries of the political are controversial. Techniques to influence public opinion have always been used when hegemonic relations and the political order are questioned and when collective action organizations seek new allies.

Spokespeople for a number of collective action organizations and a few social scientists as well as popular authors have condemned opinion formation as an activity for civil society. They reason that such campaigns are contrary to Swedish democratic practice, for they commercialize organizational goals and threaten traditional internal democracy, which has been based on support mobilization in study circles and by social integration and anchoring of decisions. They feel threatened by the SAF's ability to bypass political parties and parliament by renting media arenas to reach out directly to the Swedish public. The SAF has successfully used mass media, but so have other organizations. A case in point is the SNF, which has successfully used the mass media to mobilize support for itself and the environmental issue. With the help of the mass media, the SNF and other environmental organizations have made green values a part of postmodern Sweden.

The important question is whether critics of modern opinion formation believe that collective action organizations should play a role in political debates. Are they in principle against attempts by collective action organizations to exercise influence over politics? Would they also have criticized attempts by Swedish unions to gain political influence through the use of opinion formation techniques– strikes, demonstrations, newspaper publication, and 'oral propaganda'– which were modern at the time? Closer scrutiny of their arguments shows that the organizations and writers most critical to this development and in particular the behavior of the SAF fear that opinion formation will lead to a loss of political influence for those organizations favored by social democratic political hegemony. Collective action organizations view media campaigns and other efforts to create favorable public opinion as important means to influence politics. This is why they invest in them. The key issue for political democracy is whether economic resources should determine political influence. It is possible to regulate lobbying and opinion formation activities. Lobbyists in the U.S. must be registered, and contributions to political parties and candidates are limited by law. An issue for debate is whether the Swedish government should consider such regulation.

A final important element of the ability of collective action organizations to form public opinion and influence politics is the presence or absence of powerful allies. Leaders are also chosen for their ability to attract and shape effective coalitions of political actors who will further the organization and its cause. Both their communicative and coalition-building skills help frame and organize the activities and goals of collective action organizations by rendering them meaningful and relevant to potential allies and the general public. Leaders play an important role in defining organizational policy and profile for outsiders.

Coalitions are networks which are created for different but definite purposes. Numerous coalitions have been built in the six periods studied in this book. A prominent one is opinion coalitions. Examples include the cooperation between the labor movement and liberals in the struggle for and against universal suffrage and those which formed in the referendum campaigns for and against prohibition, nuclear power, and EU membership. Many opinion coalitions are short-lived because they unite very diverse and frequently conflicting groups in a struggle that is usually won or lost in a rather short time span. A second kind is interest coalitions or cooperation among organizations representing similar interests. The main associations of organized business– the SAF and IF– have cooperated with each other in various ways over the years, and a merger has even been considered; the LO and TCO developed in the 1970s what they called a wage earner front and are members of the Nordic Council of Trade Unions (NFS); farmer cooperative and union organizations cooperated and finally merged in the early 1970s, and contact among the LO, PRO, and ABF has intensified in the 1990s.

The history of civil society is full of examples of cooperation among collective action organizations representing the same social class. Organizations with members from the working class have formed strong and stable bonds. Their contact has been an important pillar of social democratic political hegemony. Other kinds of interest coalitions have become more prominent since the late 1950s. They involve such postmaterial issues as peace, opposition to nuclear power, environmentalism, human rights, and new diseases, e.g., AIDS. New organizations of the volunteering sector and NIMBY issues tend to engage people from different political camps and social classes. Collective action organizations have also begun to cluster around the issues of consumerism, defense of the Swedish welfare state against proposals for cutbacks, and women's interests. The networks created by these groups tend to cut across class lines and involve large and important as well as small and obscure organizations.

Swedish networks have traditionally involved stable contacts between collective action organizations and government. Government has generally been a very powerful ally of important collective action organizations. It has supported organizations in various ways and, as shown in chapter 5, shared political power with them. Contact has been particularly frequent and intense with the ministries dealing with specific public policy sectors such as agriculture, industry, health and social affairs, labor, transportation, public administration, education, and environment. Their contacts has led social scientists to call them 'embassies of special interests' (Petersson, 1989, 121, my translation).

Collective action organizations have created coalitions with political parties throughout most of Swedish civil society history. This is a fourth kind of powerful ally. The network between the Social Democratic Party and LO and Center Party and LRF have impacted politics greatly. The Social Democratic Party has been very reliant on financial contribution from the LO. Farmer organizations stopped contributing money directly to the Agrarian Party in the mid-1940s. Associations of organized business have had friendly relations with the Liberal and Conservative Parties, and organized business contributed financially to Conservative Party election coffers until the mid-1970s. Organizations have helped political parties in nonmonetary ways as well. It has been common for them to write parliamentary motions for individual MPs, participate in internal party policy groups, and help parties get media attention for issues of mutual concern. The Liberal Party has traditionally had close ties with the free church and teetotaling movements; many of its MPs are movement members and leaders.

Collective action organizations have even sought to disassociate themselves from political parties. The TCO exerted some effort in the 1980s to explain that it was not part of the social democratic network of political hegemony. A more recent example comes from the Swedish Federation of Free Enterprise which has attempted to convince journalists, the general public, and political parties that it is not an ally of New Democracy, the populist party that was in Parliament from 1991 to 1994. Critics of the position taken by collective action organizations on particular issues have attempted to mobilize support against them by labeling them as associates of a political party which a large proportion of the organizational membership does not support. This happened to the TCO in the 1970s and 1980s.

Collective action organizations generally sought contact with the labor movement in the second, third, and fourth periods. Do collective action organization still find this contact beneficial or will they try to disassociate themselves from the party, due to the erosion of social democratic hegemony? Many powerful alliances have been reconsidered in later years. Change in political opportunity structure and increased voter mobility has led collective action organizations to evaluate the pro and cons of identification with one political party or parliamentary bloc. Examples come from the three pillars of the Swedish model– organized business, farmers, and workers– and the environmental movement.

The SAF began in the 1970s to wonder whether non-socialist governments and parliamentary majorities really made a difference politically. It decided to become an initiator of political change by bypassing the Conservative and Liberal Parties and appealing directly to the Swedish public through opinion formation. Its goal was to change the Swedish system of government, weaken the influence of the peak unions, and erode the hegemony of social democracy. Its choice seems to have been a wise one. The SAF is now an independent force in Swedish politics.

The LRF began to establish contacts with other political parties in the 1980s in its attempt to renew its organizational structure and policy. Dissatisfaction with government agricultural policy, growing awareness of the declining importance of farmers as the basis of the Center Party, and the impossibility of a new government coalition between the Center and Social Democratic Parties in the near future led it to decide to lobby all parliamentary parties. Yet no matter how much money it puts into opinion formation, it has not become an independent force in Swedish politics and remains reliant on the Center and Conservative Parties for support on agricultural issues in parliament.

Collective affiliation of LO members to the Social Democratic Party was a topic of communist and non-socialist as well as public and parliamentary debate for decades. Pressure from the political environment, internal criticism within both the LO and Social Democratic Party, and escalation of the 'war of the roses' led to a change of policy (Sainsbury, 1991). The LO is still reliant on the Social Democratic Party, and it remains to be seen whether this dependence will lessen if LO member increasingly vote for other political parties. Party-union contact is still valued highly in union circles, and most union leaders are social democratic supporters (SOU, 1990, no. 44, figure 9.9).

Finally, disappointment with the nuclear power referendum outcome led the environmental movement to invest less in networking political parties. The modern Swedish environmental movement puts more effort in courting the mass media and general public. Different strategies are used. Greenpeace creates events to rally public support while quietly discussing environmental policy with the Ministry of Environment and National Environmental Protection Agency. The SNF mobilizes support for the environmental cause by grassroot actions, which aim at raising the consciousness of Swedish consumers to purchase green products, and publicizes at present the best interest group rating of the parliamentary parties of any collective action organization. Other, smaller environmental organizations use civil disobedience to create events and criticize governmental and business actions which they consider as anti-environmental.

Changes here have generally signified that collective action organizations invest more time in opinion formation than party contacts. Organizations also play a more conspicuous role in election campaigns. Among their tactics are interest group ratings and advertisement campaigns.

The parliament has provided another setting for contact between collective action organizations and political parties. Smaller organizations have always lobbied members of parliament, and larger ones have begun to do likewise. They send them letters and help MPs write motions on issues of importance to the organizations. The Conservative Party has found business associations quite supportive in this regard. The FBF has helped women MPs practice debate techniques. Certain organizations have been able to create network groups for members who are also legislators. As discussed in chapter 2, parliamentary groups of supporters of collective action organizations have existed since the early 1900s. A group of supporters of Amnesty International was created recently. What is particularly interesting about these groups is that they have attracted members from the different political parties and even across political blocs. Women MPs began to discuss the issues of women across party and bloc lines after the 1991 election.

Granting legitimacy to collective action organizations

Governments grant legitimacy to collective action organizations through laws, economic patronage, and accessibility to the political process. Legitimacy can be given or withheld, and there are numerous instances of how government has used its power to favor certain organizations over others. Regulation of voluntary associations– i.e., social movements and interest organizations– has been a political agenda item in various periods. Yet no special law has been enacted to limit their autonomy. Swedish politicians debated in the early years of this century whether voluntary associations should be forced to register with government to be classified as legal persons. Parliament debated regulating their internal arrangements. At least three government bills proposing regulations were voted down in earlier decades. The legal issues were freedom of association, right of assembly, legal standing of voluntary associations, need for regulation to check the increasing importance of collective action organizations as a political force, and rights of individual members and minority groups within them.

The most serious regulation attempt came early in the second period, when a government commission submitted a detailed forty-five paragraph long law proposal with provision for collective action organization legal status and regulation of internal arrangements, i.e., stipulations on formulation of goals, conditions for membership, member dues, composition of governing board, organizational decisionmaking procedures, auditing procedures, organizational size, etc. The proposal prohibited voluntary organizations from engaging in economic activity. It infringed on organizational autonomy, and the commission knew that it would be difficult to implement without endangering the right of association. Therefore, it proposed that this right be extended to all kinds of activity as long as it did not harm Swedish democracy. This meant that organizations promoting violence were declared illegal. This limitation was codified in 1934, when national socialist and fascist groups emerged and is now written into the Swedish constitution. The commission even discussed legislation to target labor unions, but its members believed that such a measure would be viewed as 'detested class legislation' (see *Förslag...*, 1903, p. 49, my translation).

The political debate on the proposal clearly was a partisan one. The more conservative and traditional Upper Chamber of Parliament generally approved regulation, while the Lower Chamber, with members with roots in Sweden's first collective action organizations, considered it only necessary to regulate economic associations. The Liberal and Social Democratic Parties opposed the proposal. The Conservative Party believed that government control was necessary because collective action organizations were becoming powerful. Conservative arguments were carefully phrased, but the party was particularly worried about the strength of labor market organizations, especially unions. The proposal was not passed in parliament.

The debate on state control of civil society was interrupted by World War I and laid abeyant during social democratic rule in the 1920s. Government adamancy on dominating civil society gradually subsided. The debate ended in the mid-1950s when a new law on economic associations– i.e., cooperatives– was passed in Parliament. The issue became less urgent once collective action organizations became institutionalized and the labor movement gained a hegemonic hold on politics. Also, collective action organizations had by this late date shown themselves to be responsible political actors: they were, as Swedes say, housebroken. A political scientist summed up the decades-long debate: 'Swedish civil society has successfully opposed legislative attempts, but through its own strength it has– in cooperation with courts–

also filled at least part of the goals of the legislation' (Heckscher, 1951, p. 119, my translation).

Collective action organizations are granted legal capacity without official registration. Criteria for status as legal persons are minimal. They are required to have a governing board and a constitution, which defines the organizational goals. They must be institutionalized to an undefined degree, include at least five persons as members, abide by publicly acceptable customs, treat all members equally, follow the principles of representative democracy, and show allegiance to the laws of the land. Certain collective action organizations are regulated more strictly. They include cooperatives and other organizations, which are delegated public authority for specified tasks and which seek tax exemptions and government patronage. Though not codified, many law experts argue that these organizations must follow a chain of decisionmaking that is hierarchical, based on representative democracy, and possibly also reflective of the democratic practices that guide local governmental decision-making bodies.

Freedom of association is so embedded in Swedish democracy that government is very hesitant about legislating against unwanted associations. For instance, a United Nations (UN) convention compels its signers to enact laws which prohibit racist organizations and make membership in them a criminal act. Sweden ratified the convention in 1965 but has not legislated on the issue. The question has renewed currency in the 1990s. In general, government believes that collective action organizations– given the exceptions just mentioned– should be capable of governing themselves in a publicly responsible way. This reasoning has even applied to labor market organizations. Government investigations in this field usually encourage the unions and employers' associations to exercise self-regulation and assume public responsibility for their actions. Significant and direct government involvement in labor market relations has been viewed as an extraordinary measure. Nevertheless, the fact that Sweden has become increasingly integrated into the international economy has led government to closely monitor collective bargaining sessions, meet frequently with labor market leaders, and use income policy measures to reach satisfactory results.

Public debate on the political influence of collective action organizations began again in the mid-1960s and became a controversial issue in the 1970s and 1980s, when conservative forces questioned the role that these organization play in the political process. A very vocal critic was the Swedish Civil Rights' Movement (MRR) whose two basic issues have been Swedish corporatism and rights of citizens as members and non-members of collective action organizations. A central topic of debate is 'citizen right of nonassociation'. Proponents criticized collective affiliation and the political influence of corporatist collective action organizations, which was said to force individuals to become members. Social democrats have generally viewed criticism here as an attack on wage earner solidarity, virtually impossible to guarantee constitutionally, and unnecessary as the constitution already prohibits compulsory membership in political, religious, and certain other organizations. All non-socialist parties along with the Environmental Green Party have argued that the right was necessary because unions are not representative of their members, their internal democratic arrangements unsatisfactory, and the right of nonassociation and right to exit union organizations not guaranteed sufficiently. An important argument has been a UN statement on human rights in which the right of association and non-association are seen as irrefutable rights of citizenship. A commission appointed by the non-socialist government to investigate issues of individual freedoms believed that this right could not be satisfactorily regulated constitutionally.

Labor law is another topic of controversy. Particularly the Conservative Party believes that current law favors unions over employers' associations. The non-socialist government appointed a commission to evaluate labor law. Internationalization and individualization of Sweden were emphasized in the committee instructions. Many of the ideas included in its instructions were taken directly from a detailed report written by the SAF. Social democrats as well as union representatives believed that the non-socialist government was engaged in union busting. Yet they seem to forget that the social democratic government took several initiatives in this field while in office. It appointed a commission to review labor law that was later replaced by the one appointed by the non-socialist government. Also, it was highly involved in reforming the ground rules for collective bargaining, and the report from its committee here was severely criticized by the LO for favoring employers over unions. It is too early to evaluate how the new social democratic government will deal with labor law.

Swedish laws and legal practices have generally integrated collective action organizations in the political process. This is, in fact, constitutionally guaranteed. The right of collective action organizations to participate in the system of official comments and be delegated public tasks are written into the Swedish constitution. Government has even created forums for the integration of collective action organizations in the political process. This has occurred for senior citizens and in the agricultural sector. Membership of collective action organizations on lay boards of public agencies was codified in 1986. This right has also debated. Problems of governability led many political actors to question whether corporatism still benefitted Sweden. This occurred in the fifth and sixth periods. Collective action organizations were removed from agency lay boards in 1992, and questions have been raised about the value of comments by collective action organizations on government proposals.

The final crucial way in which government grants legitimacy to collective action organizations is through government patronage, i.e., economic grants from government to collective action organizations. Government patronage involves very complex and confusing practices and principles. It is unfortunate that few scholars have attempted to understand it, for it deals with the central dilemma of political democracy: the dividing line between civil society and government and government control of collective action organizations. All levels of Swedish government are involved in funding free church, adult study, sports', teetotaling, women's, cultural, environmental, handicap, hobby, humanitarian, pensioner, civil defense, youth, handicap, and union organizations. The tendency until the sixth period has been for government to extend its financial commitment. Today financial problems are forcing reevaluation of this economic and moral commitment to collective action organizations.

Concern about government control of grant usage began when collective action organizations started to receive economic support in the mid-1800s. It was an important topic in the first and second periods, a time when parliament also debated the need to regulate and control collective action organizations. This same concern is again important in the sixth period and is another example of the interesting similarities between the early and late periods of civil society history.

Individual MPs began in the first period to submit motions for funding of the activities of those organizations with which they sympathized, and organizations requested financial support directly from the government for special tasks and projects. Various voluntary associations– including the SNF, FBF, and peace movement organizations– received funding. Piecemeal financial assistance became very common. It augmented membership fees and contributions from other patrons. Grants were given for activities performed by organizations– not to the organizations themselves– and were seen as official and public acknowledgment of the public importance of their activities. Initial important arguments for allocating tax money were the lack of suitable government bodies to perform specified public service tasks and costliness of creating such bodies. These arguments gradually became an implicit principle for government patronage and would in all periods be used in parliamentary and public debates.

A major consequence has been that grants to organizations to perform services have not always reflected their goals. The best example here is grants to teetotaling organizations to inform the public about the dangers of alcohol but not to mobilize support for prohibition, which is the basic reason for their existence. Government grants for this public service began in the mid-1800s. Teetotaling organizations are, with few exceptions, the only kind of collective action organization to receive such funding. Many organizations– including unions and the Verdandi– are also dedicated to lowering the consumption of alcohol, but they cannot receive grants, as they are not classified as teetotaling societies because they do not require a vow of abstinence from their members.

General principles for government patronage have never been formulated systematically. Principles has been discussed in parliamentary debates on financial support for particular organizations. The generous funding of the sports' movement, which received its first annual grant in 1913 after years of futile attempts, and the Swedish Rifle Association, an association funded for the first time in 1863 for its contribution to national military defense, have repeatedly been compared with the requests for peace, teetotaling, and adult study associations. Here the argument is that if one organization receives money others should as well. Later this practice would generally impede government from cutting grants for certain associations while increasing them for others.

Criticism of government patronage was voiced in the parliamentary debates in the first two periods. Critics argued that government should not allocate money without knowing exactly how it would be used. They were concerned about the accountability and trustworthiness of collective action organizations. Several parliamentary debates in these early years even involved detailed review of the economic records of funded voluntary associations. The teetotaling movement was targeted here. Critics even questioned whether organizations that requested grants for specific purposes could perform the tasks properly. Parliamentarians asked in a debate in 1929 if the SFSF and the Swedish branch of the Women's International League for Peace and Freedom could really perform the public task of distributing war prevention information because they were in fact against military defense. This same question could have been asked about money appropriated to the teetotaling movement for information on the effects of alcohol. Skeptics also asked whether it was a good idea to let the government rather than parliament assume responsibility for distribution of funds. Many MPs feared that parliament would play a minor role here and that the government would not allocate money in an objectively and fairly. They emphasized the need for principles or criteria for the administration of government patronage. These issues would be raised again in the sixth period.

Opposition to government patronage gradually died out. Government had, by the end of World War II, made a large financial commitment to Swedish civil society. Funding was still granted for activities performed by voluntary associations, but it also became customary to allocate financial support for organizational maintenance. A government commission openly declared in 1952 that taxpayers' money should be allocated to help teetotaling associations solve their economic and organizational problems and recruit new and hold on to old members. Funds were also openly allocated for the administration of Swedish adult study associations.

The development of government patronage has involved standardization of procedures in the early postwar years and then decentralization of its administration to coordinating units either legitimated or created for this task in the 1970s and 1980s. All reform attempts have had administrative simplification and fairness of allocation as primary goals. Past reforms have not solved basic problems. Today the trend is for central government authorities to play a minor role and establish clearer general principles for grant allocation, a development reflective of other public administrative reform efforts.

Government boards were created by ordinances in the late 1970s to administer patronage to the free church and teetotaling movement. Board members come from the movements themselves. The government confirms their nomination. Frequently politicians are appointed as chairpersons. A granting board for adult education associations was established in 1991. Its members come from adult education associations as well as the Federation of County Councils. Granting boards have not been created for all sectors of civil society. Certain public agencies and government ministries are responsible for distributing grants, as in the case of handicap organizations (funded by the Ministry of Health and Social Affairs and National Board of Health and Welfare), youth associations (funded by the State Youth Council), women's organizations (funded by the Ministry of Public Administration), and immigrant organizations (funded by the Swedish Immigration Board). The Supreme Commander of the Armed Forces is responsible for distributing most grants to civil defense associations. A different model is used for sports' associations. The government has delegated this authority to the peak association the Swedish Sports' Confederation.

Voluntary associations even pool their resources to apply for patronage. An association was formed in 1936 to coordinate grant applications from adult education societies, and one that was initially created in 1972 to distribute information on foreign aid began, after encouragement from the Swedish International Development Authority, to administer grants for foreign aid projects to collective action organizations in 1992. It is called the NGO Foreign Aid Information Center (BIFO). Most ministries include grants to voluntary organizations in their budgetary requests, which are incorporated into the annual budget bill.

A serious problem with government patronage is the lack of an adequate definition of what is meant by voluntary associations. Several different definitional criteria are offered in government papers. The criteria most often cited are national geographic breadth, internal democracy, membership open to all citizens or categories of citizens, and national membership of at least 1,000 or

3,000 members. However, the criteria are not applied systematically. Handicap and immigrant associations need only 500 and 1,500 members respectively to receive funding, and there is no specification that free churches must be governed democratically.

The criteria of democratic governance has been applied to deny the environmental organization Greenpeace status as a voluntary association. Here democratic governance refers to representative traditions of established Swedish collective action organizations, i.e., congresses to elect executive boards. The problem is that Greenpeace fulfills certain other criteria much better than many established associations. Its checkbook membership is much larger and geographically widespread, and it promotes debate on a very important issue of postmodern Sweden. Focus on congresses as democratic decisionmaking bodies is problematic in itself as there is debate on their value within older voluntary associations. A few unions have even decided to discontinue with the practice of holding congresses, as they are proving to be expensive, inefficient, and inflexible decisionmaking methods. Greenpeace has not sought government patronage but cannot receive tax exemptions, as it is not classified as a voluntary association. It can be questioned whether choice of criteria has always been objective, and it seems that certain criteria are given higher priority than others.

Frequently the government specifies in ordinances which organizations are eligible for patronage, as in the case of the free church, teetotaling, immigrant, women's, and adult education associations. The government decides which collective action organizations are worthy of government patronage and, thus, defines which are *legitimate* free church, teetotaling, and adult education associations. This practice has not been completely problem-free. Certain ministerial units responsible here have not formulated specific criteria on which to base their decisions, and it is not always necessary for them to motivate why they deny legitimacy to particular associations. In many cases, associations cannot appeal decisions made by the ministerial unit. The Commission for State Grants to Religious Communities is a case in point.

Moreover, it is not uncommon that the legitimating bodies apply different criteria. Some units will deny government legitimacy to voluntary associations on the basis of their small membership size or narrow geographic breadth, as has been the case with certain teetotaling organizations, while other bodies deemphasize these criteria and emphasize the democratic traditions of the organization. A few free church associations have not been legitimated because the responsible government unit has deemed their activities as destructive to society. This criterion has not been defined officially, but it is clear that it concerns activity considered as inappropriate according to contemporary Swedish norms. It would seem that many of the implicit rules which are applied to give collective action organizations legitimacy reflect a political correctness that preserves the status quo and glorifies the achievements and character of the collective action organizations that emerged in the first three periods.

Another problem with government patronage involves organizational categories, an issue also raised in the early decades of Swedish civil society history. For example, youth associations must seek legitimacy from the State Youth Council in order to receive government grants. Regional and many local governments use this stamp of approval as a basic criterion for their own funding decisions. The problem is that the State Youth Council has not formulated a clear definition of what is meant by youth associations.

Few political actors have viewed the lack of clear principles for granting government patronage as problematic. The system has not been considered arbitrary or unfair, and there are few– if any– instances of blatantly unfair treatment. An important explanation is government generosity in funding decisions and general agreement on what is considered as fair and objective treatment of collective action organizations by government. The strong society and later Swedish model ensured this development. There are signs that this viewpoint is now being reconsidered.

Yet the absence of clear rules and principles is problematic. The lack of a clear definition of a voluntary association has created confusion, and funds have been allocated to organizations not strictly voluntary in nature. A recent report clearly showed this to be the case in the area of peace work (UD, 1992). Also, distribution of government patronage has become routinized, and cliental relations have developed between granting agencies and grant-receiving associations. Granting bodies have tended to cater to those organizations with which they have stable ties. This excludes newer organizations. One potentially dangerous consequence is that government, through its funding decisions, inhibits the development of civil society by favoring established collective action organizations. Unclear principles and the tendency to formulate diffuse goals have led voluntary associations to wonder what government really wants of them. Allocation principles are more important now that economic cutbacks lead to more competition over funds.

Certain voluntary associations refuse to apply for financial support for they believe that government will use it to force them into passivity. The National Association for a Drug-Free Society (RNS) believes this (see Wikström, 1992). The Swedish branch of Amnesty International and Guardian Angels do not accept any government patronage. Free churches have had the most lively debates on the advantages and disadvantages of funding. Spokespersons for patronage have argued that 'the free churches should not...be excluded from benefits received by other voluntary associations'. Opponents feared that economic reliance on government would lead free churchmen and -women 'to remain silent about societal problems which government was responsible to rectify' (FS, 1970, pp. 19, 26, my translation). Economic difficulties led the free churches to view government patronage in a more favorable light, and societal developments forced them to accept it. Annual as well as targeted grants are now accepted, and the free church movement actively monitors and compares its level of patronage with that received by other collective action organizations.

Since the 1930s, commissions appointed to investigate government's commitment to civil society have argued for more funding. Funding was granted in the third period to reduce organizational costs but not replace membership fees. Today considerable sums of money are allocation to maintain collective action organizations. Transfer payments are common as are grants to stimulate recruitment of members and subsidize salaries for administrative personnel. Such grants have increased over the years and accounted for more than 70 percent of the total sum of national government patronage by 1991. Funding for public services performed by voluntary associations, the reason for government grants in the first place, represents slightly less than 30 percent of the total (SAFAD, 1991). The national government allocated about 1280 million crowns (ca. 185 million dollars) in its 1992 budget to support public service efforts by voluntary associations (SOU, 1992, no. 130, appendix 6). The total amount of national government funding to voluntary associations was calculated at 3,370 million crowns (ca. 560 million dollars) in 1988. Local government contributed about 6 million dollars in 1986 or approximately 15 percent of the costs incurred by voluntary associations within their jurisdictions. Municipal funding began to increase dramatically in the 1960s (see Micheletti, 1993).

Politicians began in the early postwar years to believe that government should not be allowed to encroach on the right of association by demanding that organizations account for how they spend taxpayers' money. Over the years it has become clear that encroachment has occurred. Both the organization and activities of voluntary associations have been highly effected by government patronage. Studies of adult education associations have shown that the rules governing grants have led to structural inflexibility and difficulties in curriculum renewal. Adult education associations have been trapped by government patronage. Granting procedures have slowed down the adjustment of adult education associations to the zeitgeist.

Several voluntary associations are dependent on government patronage for their survival. Many free churches and women's movement organizations as well as smaller teetotaling and social political associations would not exist today were it not for government patronage (see Wikström, 1992, Fi Ds, 1992, no. 114). The SFSF's dependence on government increased from 6 percent in 1978 to 33 percent in 1990 (Kron, 1991, p. 69). Membership fees accounted for less than 10 percent and government patronage for over 40 percent of the 1990 budget of the Save the Children Foundation (*Barnen & Vi*, 1990, Nn. 4). A study from the early 1980s showed that about 77 percent of the finances for national, peak immigrant organizations came from the national government (Bäck, 1983, p. 51). This led a scholar of immigration to conclude: 'the immigrants' central organizations are no way dependent upon their own constituencies: they subsist on state funds and do not give a damn about what their members think' (quoted in Schierup, 1989, p. 247). Even such an established, large, and respected association as the Swedish Red Cross is highly indebted to government. Membership fees accounted for only 6 percent of its total costs in 1986, which represents a significant change from earlier decades.

Membership fees for most Swedish voluntary associations, with the exception of unions, are generally quite low– between 100 and 200 SKR per year. Large memberships are important for the calculation of government patronage, as larger associations tend to receive more support than smaller ones. One wonders whether organizations keep their membership fees low to retain uncommitted members in order to receive higher levels of government funding. Membership records are, in certain cases, ill-kept and include names of deceased members.

Complicated and bureaucratic granting procedures, diffuse principles and definitions, and vaguely-formulated government goals make it extremely difficult to evaluate the effects of government patronage. The issue of accountability is once again a political concern. Various public bodies, including a number of government commissions, have studied government patronage. All investigations propose reform. Parliament has debated the issue in the fifth and sixth periods. The National Audit Board (RRV) and Agency for Administrative Development (SAFAD) have critically evaluated government patronage and propose better management of national government funds. Several voluntary associations have also assessed the advantages and disadvantages of government patronage; adult study associations have been particularly active in this regard. A public debate on this issue began in the 1980s.

Recent government reports have concentrated on the adult education, youth, peace, and women's associations. The main issues are: (1) level of government financial commitment; (2) need for clear principles, fair granting procedures, satisfactory evaluative methods; (3) dependence of civil society on government; (4) role of civil society in the current age of transition, and (5) need for grants to stimulate stagnating organizations. The non-socialist government declared that it wanted to channel more money to the volunteering sector, and it appointed a special commission to evaluate government patronage and find ways to simplify the administration of grants and make collective action organizations less dependent on government. It has drastically cut funds to labor unions and adult study associations. These issues were broached by the social democratic governments of the 1980s and RGSM in the 1970s. The title of a government commission on youth associations 'More for Less' sums up well the current reasoning on government patronage. It will be interesting to follow the new government's approach to these issues.

Women's and youth associations as well as volunteering associations are targeted in the sixth period. Voluntary associations are targeted because they are seen as an alternative to public sector involvement. Parliament decided in 1982 that women's associations should receive sufficient annual funding to plan their activities without worrying about financing them. The reasons for patronage are classic:

> By giving patronage national government wanted to show that it considered women's associations to play an important role as agents to improve the conditions of women in society and alleviate the differences between men and women. By starting a system of patronage government also wanted to reach equity with other organizations already receiving funding (Fi Ds, 1992, pp. 8-9, my translation).

The general trend is that all levels of government are attempting to decrease their funding levels, and they are reconsidering their priorities. Many voluntary associations are affected negatively by this. The age of transition is proving particularly difficult for adult education associations and union organizations. New tax laws have also increased the costs of maintaining collective action organizations, and EU membership may affect tax exemptions.

Many collective action organizations believe that government is treating them unfairly. They have once again begun to lobby government and mobilizing support for piecemeal funding. Their basic arguments for increased patronage are: comparisons with their higher level of funding in the past and with other organizations to show that they are victims of discrimination; the need of government to give priority to the service performed by them. Organizations also put pressure on the political parties with whom they have friendly relations. Free churches and peace associations lobby the Liberal Party, and unions contact the Social Democratic Party. These efforts are often successful.

Certain organizations see grant cuts as motivated by partisan politics. Adult education societies have collected statistics to show that non-socialist majority local governments discriminate against cultural associations. The president of the Verdandi has stated that his organization is stigmatized by social democratic local governments, for they fear criticism were they to grant funds to voluntary organizations with friendly ties with the party (see Wikström, 1993). Labor unions felt targeted by the non-socialist government, which had successfully proposed legislation to cut grants to union courses on the law of co-determination (MBL) by two-thirds and curtail tax reductions for people who pay union dues. Yet the question is whether a social democratic government would have acted much differently, as it was a social democratic government that commissioned the evaluation of grants to union education.

Government has even taken the initiative to help collective action organizations find other sponsors. A reform of gambling laws is planned to make it easier for voluntary associations to generate economic resources from lottery sales, bingo, casinos, etc. Three government commissions have investigated this issue in the 1980s and 1990s. Lotteries were an important source of income for voluntary associations in the early periods, until government began seriously to compete with them. Another proposal is tax deduction for citizen contributions to civil society organizations.

Internationalization, globalization, and intermesticity

Relations between national and supranational institutions have been a part of politics in all countries in the twentieth century. The decision by the Swedish government to apply for EU membership and the creation of the internal market within it have made contemporary Swedes more conscious of their attachments to the rest of Europe. There is even growing general awareness that inter-governmental cooperation is the only solution to many serious problems facing the world today. Globalization of financial markets has forced Sweden to realize how interdependence effects her domestic welfare. These developments are affecting perceptions of Swedish national identity, and they are changing the Swedish political opportunity structure.

Internationalization is not a new development. Swedish historians use the term to characterize the political situation of Sweden in the 1800s. A great number of international collective action organizations emerged at this time. Important reasons are development of extensive interconnections among nation-states and technological developments. Free trade agreements in the mid-1800s mark the first wave of international economic integration (see Stråth, 1992). Skilled craftsmen traveled abroad in search of work in these years; many of them spread the ideas and ideology of emerging collective action organizations. People began to feel solidarity across national boundaries, as evidenced by the spread of collective action organizations throughout Europe and North America. Good examples come from the religious and labor movements. Such technological advances as railroads, telephones, and the telegraph made contact across national borders feasible.

As shown in earlier chapters, many of Sweden's most respected collective action organizations were imports. Later these organizations became active in defining the Swedish national identity. Nationalism and national identity in Sweden as elsewhere grew stronger in the 1900s. Economic protectionism, political developments, and wars encouraged countries into isolationism. Barriers to free movement of labor began to reappear in the early 1900s. The creation of unemployment benefit societies eliminated the need to solve the problem of unemployment by seeking work in other countries. General restrictions on international labor mobility were imposed in Europe after World War I. Nations guarded themselves against foreigners by requiring them to procure passports, visas, and work permits. Until 1914 it was possible for people to travel freely within Europe; passports were introduced in Sweden in 1917.

The number of international governmental as well as nongovernmental organizations grew rapidly after World Wars I and II. People also became more nationalistic, but many of them believed that new institutional global arrangements were necessary for the promotion of peace and solidarity. Technological advances, creation of institutions to promote peace, and advancement of humanitarian goals for social welfare and human rights have led to the establishment of international governmental (IGOs) and nongovernmental organizations (INGOs). The idea behind these endeavors is global democracy.

INGOs emerge earlier than IGOs. They mobilize support to persuade national governments and IGOs to take action on specific and general problems. INGOs are also numerous. In many ways, they play the same role as collective action organizations in national settings, as they desire to change the boundary of the political at the global level. They form the basis of the global civil society, which began to develop more rapidly after World War II.

Swedish collective action organizations have always been engaged in international activities. As stressed earlier, traditional and many newer collective action organizations, which began to develop in the 1960s, have roots in international movements. Swedish government encourages their involvement in international settings through grants and official declarations. Today global problems, changes in the international political setting, and transformation of the Swedish political opportunity structure make international activities more important for collective action organizations. They are active internationally in various ways. Some organizations are highly involved in INGOs because they want to exercise influence over decisionmaking. Trade union involvement in the European Trade Union Confederation (ETUC), the umbrella organization for European national unions, is one example. Other organizations want to help citizens in foreign countries through missionary work, foreign aid services, and contact with sister organizations. Free church, teetotaling, union, and citizen rights associations are involved in this kind of international work. An increasing proportion of Swedish foreign aid is channeled through collective action organizations. A third group, as exemplified by such humanitarian associations as the Church of Sweden Aid, Swedish Red Cross, and Amnesty International, mobilizes support for international causes in Sweden.

All organizational levels are involved in international activities. More resourceful organizations have their own international secretaries and units. Less resourceful ones find it difficult to develop international contacts fully. It has also been common for collective action organizations to coordinate their international activities by establishing cooperative networks. Examples include the Swedish Co-operative Center (SCC), which coordinates the KF's and LRF's foreign aid efforts in Africa, and a LO-TCO board, which administers financial aid from the blue- and white-collar movements to unions in less developed countries.

International activity involves various global levels. Swedish collective action organizations participate in Nordic, Baltic, European, and global settings. The newest contacts are with the Baltic and former Eastern European countries. Organizations as diverse as trade unions and the RFSL help similar groups in these countries establish themselves. The general aim is the creation of civil society. The SNF participates in a network called Coalition Baltic Clean Up, which involves several environmental organizations from countries in the Baltic Sea area.

Nordic cooperation among collective action organizations is the best developed in Sweden. Most if not all Swedish organizations are members of at least one Nordic organization. An investigation from the mid-1970s reported over 400 such organizations. Business associations accounted for about one-third, union organizations for roughly 40 percent, and the remainder were sports, youth, leisure time, religious, welfare, and humanitarian associations (Hallenstvedt, 1974). Most likely the number of Nordic organizations is higher today. Resources have not been ill-spent on Nordic networks, as contacts have dispersed ideas on a Nordic basis (Karvonen, 1981). Swedish and Norwegian collective action organizations have been the most Nordic-oriented in the postwar years (Schiller, 1985). The reason is that Sweden and Norway considered Nordic cooperation as a viable alternative to membership in the European Economic Community (EEC). EEC membership was politically controversial in both countries.

Internationalization again became a very important topic for all established Swedish collective action organizations in the sixth period. Current economic developments, in particular the growing importance of multinational corporations and their impact on national autonomy, deregulation that cut back government control and opened markets to domestic and foreign entrants, the Single European Act or legislative framework for the internal market, and the eventuality of Swedish EU membership have created a situation somewhat reminiscent of the one characteristic of the late 1800s. These developments are leading to a new sense of urgency for international contacts and cooperation. As in the past, an important reason is international economic integration, which implies a freer economic market as well as less protection against importation of foreign labor and exportation of capital and employment opportunities.

Internationalization impacts Swedish trade unions greater than other collective action organization. Unions became politically powerful in the period of the Swedish model. Swedish corporatism and social democratic political hegemony favored them and encouraged unionists to identify more with the Swedish nation than the international labor movement (see Logue, 1980). Corporatism was a viable alternative to internationalism for unions and other collective action organizations. Unions were protected by the political opportunity structure of the Swedish model. Corporatism and strong, self-resourceful centralized government complemented each other well in Sweden. For these reasons, international contacts in the years of the Swedish model was more concerned with foreign policy, which promoted Swedish neutrality, than domestic economic and welfare state policy.

Now unions are undergoing a crisis, which can be explained by the emergence of transnational forms of representation and regulation as well as changes within Sweden. Decorporatization and trade liberalization are forcing labor market organizations to find other channels to exercise influence and build political networks. As in other countries, business and employers' associations were the first collective action organizations to sense the importance of new internationalism to further their economic and political interests. The international orientation of business generally encourages collective action across national borders. In fact, the new wave of internationalization involves many national and pan-European interests organizations for business (Greenwood, Grote and Ronit, 1992). New internationalism concerns European economic and political integration.

Many unions believe that economic integration can be arrested and have difficulty in seeing how it can be used to their benefit. Other collective action organizations are active in EU discussions. The views of farmers' associations on Swedish membership began to change when the social democratic government decided to deregulate agriculture in the 1980s. EU membership was seen as a way to impede this. Non-class based and non-gender collective action organizations have generally perceived the eventuality of Swedish EU membership as less problematic. This is explained by the technical and economic orientation of the EU. There are exceptions to this orientation, as shown by the emergence of the Euro environmental movement.

Collective action organizations realize that it is important for them to take part in EU policy negotiations. Many of them have begun to closely monitor

EU decisionmaking bodies. Several organizations now have offices in Brussels, the EU capital. Among them are the LRF, LO, TCO, SAF, IF, and KF. Others participate in movement networks and belong to Euro collective action organizations which lobby EU institutions and officials. There are, for example, green and women's networks active in Brussels. Activity here aims at influencing the character of policies and the EU political system as well as changing the Swedish political opportunity structure to conform to EU practices or prohibiting such adaptations from taking place. An important reason for Swedish collective action organizations to join Euro groups is their ability to stay abreast of European developments. Other reasons are access to decisionmaking bodies, knowledge and expertise, and transnational alliances or networks of like-minded individuals and associations. Also, membership may ensure that Euro organizations do not take action contrary to the interests of Swedish member organizations.

Maintaining a presence in Brussels is highly important. Investments in Euro lobbying and opinion formation activities have frequently been successful, and research has shown that activities by collective action organizations make a difference (George, 1985, Kirchner, 1980, Greenwood, Grote, Ronit, 1990). The emergence and incorporation of Euro collective action organizations in the EU policymaking process is a sign of the growing importance of a 'Euro strategy' and possibly the diminishing importance of the 'national route' to Brussels, which has traditionally meant contact between national organizations and the Council of Ministers, the final decisionmaking authority on EU matters. The main target for lobbyists is the European Commission. Its officials seek contact with as wide a range of collective action organizations as possible. Their lobbyists concentrate on this institution because they can 'begin providing input to the Commission long before potential legislation surfaces as a formal Commission proposal' (Gardner, 1991, p. 30). Lobbying the European Parliament has even become more frequent in recent years, as its scope of authority broadens.

Collective action organizations have understood that Euro organizations are not as weak, fragmented, or unable to agree on meaningful common positions as generally assumed. In fact, Euro organizations have on occasion been delegated public responsibility by the European Commission to obtain consensus and anchor EU ideas and proposals among national sectoral associations. Many Swedish organizations thus realize also that they must participate in Euro organizations to influence EU policymaking. Swedish organizations are developing a Brussels strategy. Once again, business associations have taken the lead and been the most successful. But a problem for many Swedish collective action organizations is that they find EU political culture to be foreign. It is more pluralist than the Swedish one, and collective action organizations find lobbying a new and uncommon method to exercise influence.

Integration into Europe deals with economics (free trade) and politics (creation of new loyalties) (cf. Haas, 1968). It implies creation of a new Swedish national identity and political opportunity structure and transfer of political and territorial authority to the EU. The process of economic and political integration involves different sets of actors. Individual citizens, civil society, and government are involved. Official Swedish preparations for EU membership have involved more economically-oriented than public-oriented collective action organizations. Organized labor and business have been well-represented on the twenty-four work groups appointed to integrate Sweden into Europe.

Importance of social democratic hegemony

Mediation of class conflict has been the basis of modern Swedish government and social democratic political hegemony. Government has not been passive but an autonomous force that negotiates with organized interests in decision-making settings. The main political issues have concerned the allocation of material welfare among social classes and aggregation of interests in uniform, reliable collective action organizations. The ideology of Swedish social democracy and liberalism explains this political and national characteristic, as both the Liberal and Social Democratic Parties have been welfarist in orientation. Research has recently even shown that all non-socialist parties did not oppose key aspects of welfare state reform and on occasion initiated it (Uddhammar, 1993).

Two cornerstones of social democratic hegemony have been national economic growth and the success of the party at election times (cf. Pontusson, 1988). Economic growth was necessary, as an important point of departure for the negotiations among political actors has been the ability to reach non-zero sum settlements to solve socio-economic problems. Responsibility for governing Sweden was decisive for the development of social democratic political hegemony. It reassured the LO and other social democratic collective action organizations that parliamentary democracy could be used to solve class problems. Militant action, civil disobedience, and revolutions were not needed to convince political elites to focus their attention on these problems and change the boundary of the political accordingly. Reliance on parliament in turn led non-social democratic politicians and supporters to trust social democrats as parliamentary actors. Office-holding even forced the social democrats to engage in class compromise, as they often ruled in either coalition with the Agrarian Party or as a minority government. The long tenure of the party led people to regard it as the main point of departure for political discussion; the other parties were relegated to the position of minor political actors, who reacted to social democratic proposals and argued with each other over policy initiatives.

More than the other political parties, social democrats were able to mobilize support for their vision of the future and incorporate ideas with popular support into their vision. This gave their vision a practical and adaptable quality and allowed them to take credit for innovative policymaking. Incorporation of green and gender value as well as the interest of white-collar workers into the party ideology are three examples.

The social democratic vision benefitted from as well as promoted the general change from rural to industrial society in the first and second periods. The fight against hunger and poverty, integration of all social classes into politics, provision of social security for all citizens, and a greater degree of equality of social status were included in this vision. The social movements that emerged in the late 1800s also had this general dream on their agenda. They emerged and grew strong in the same period and, together with social democrats and liberals, began to define the boundary and practices of Swedish politics. Over the years, the Swedish sense of community became identified with government action and policy. Government and civil society merged into the strong society and later the Swedish model. They are two political opportunity structures which showed that civil society and state had, at the time, the capacity for common action that benefitted the vast majority of citizens. Several tasks performed by social movements were municipalized, and many collective action organizations began to stagnate. This occurred during social democratic rule.

Early on there was understanding across party lines on the value of collective action organizations. Not every organization met with the approval of all established political actors, but there was general agreement on the advantages of collective action for political democracy. Later organized solidary among citizens evolved around dedication of government to improve material welfare and the conditions of the less fortunate. All social classes were included in the welfare state project, as explained in chapter 4. Informal partnerships and arrangements reinforced the formal government authority of the Social Democratic Party and advanced its political hegemony. Social democratic norms and practices became embedded in society in this fashion.

The value orientations of Swedish citizens and collective action organizations merged with social democracy as it grew stronger. There were good economic, political, and social reasons for collective action organizations to promote and accept the social democratic underpinnings of government. Social democratic political hegemony was sustained by diverse groups, many of which did not support the party at election time. Yet such behavior strengthened the party's hegemonic hold on Sweden. Gradually these organizations adopted or tolerated the social democratic conception of collective identity.

Social democratic political hegemony involved different levels of politics. It concerned the third dimension of power, i.e., the intersubjective level of what people implicitly understand are the arrangements through which their affairs are handled. Use of language is important here. Swedish political language and behavior has reflected a social democratic, collectivist democratic ideal (cf. Petersson, 1991b, Olsen, 1990, 1991). The words 'we', 'our', and 'us' have been more important in political rhetoric than words promoting the individual. Such words have also put pressure on deviating individuals to conform to Swedish standards.

Social democratic political hegemony involved the second and first dimensions of power as well. These dimensions concern the behavioral level of politics. It is at the second level that third dimension understandings are expressed in regularized and routinized behavior which reproduces itself culturally and organizationally in politics. It concerns implementation of politics, political democracy, interest aggregation, and representation. The first dimension of power involves politics as conducted in official debates, discussions, and decisions. Social democratic successes in enacting laws and implementing welfare policies are examples of the workings of their political hegemony in the first dimension of power. The course of events in Sweden, and in particular the wartime experiences, promoted the premises and patterns for social democratic hegemony on these three levels of power. The small state status of Sweden, growing nationalism, and centralization of politics that took place in the first half of the 1900s encouraged people to unite together and consider themselves a homogeneous group.

Swedish collective identity can be likened to small group identity. Small groups need unity to accomplish their goals, maintain themselves, and develop their collective identity (cf. Cartwright and Zander, 1968). In Sweden, these needs led to the articulation of national standards for politics and society, as reflected in the metaphors of the strong society and Swedish model. Pressures toward uniformity were accepted or tolerated because they were seen as assuring progress toward national goals. Conformity has been more imperative in such collective action organizations as the LO because they have been the main bearers of social democratic political hegemony. Individuals and organizations defying the collective norms were seen as threats to the national social welfare state project.

The prevailing conceptions of democracy and interest representation played a regulatory role in society and politics. They became authoritative rules for political, economic, and social behavior. Gradually they evolved into political correctness or a normative framework for political socialization.

Central aspects here have been the prominence of class over other struggles of recognition (e.g., gender and ethnicity), universalism rather than particularism as a decisionmaking strategy, elite democratic practice, conflict avoidance, collectivism rather than individualism, and distaste for confrontational and unconventional political behavior, market capitalism, and commercialism. Uniformity, universalism, and collectivism were necessary because particularistic identifications were seen as threatening to weaken the ability to develop the welfare state and Swedish national identity. Uniformity, universalism, collectivism and group cohesion or solidarity were the cement of Sweden. They were the characteristics that united the diverse elements of Sweden. Scholars have argued that an important ingredient here has been the desire to centralize activity in order to control it (cf. Elster, 1989).

The course of events in Sweden transformed in the fifth and sixth periods. The changes are no longer promoting the premises and patterns for social democratic hegemony on the three levels of politics discussed above. Competition over resources is causing a reevaluation of Swedish national identity and reform of the political opportunity structure. Economic problems, public sector inefficiency, war of the roses, defeat of the social democrats in local government elections and in the 1976 and 1991 parliamentary elections as well as a series of scandals within the social democratic sphere helped erode their political hegemony. Viable alternatives to social democratic thinking have been formulated in later years; the non-socialist parties have been more united, and reaction of social democrats to events in the fifth and sixth periods showed that the party was losing its grip on politics. These developments show that social democratic political hegemony grew at least in part out of divisiveness of the opposition rather than just the monolithic power of the labor movement.

A definite shift in the coalitions that formed the basis of the Swedish model is taking place, and networks are realigning. New identities are replacing the collective identity that characterized much of civil society history as analyzed in this book. Its source in such concerns as individualization, decentralization, marketization, and internationalization (Petersson, 1991b). Rather than considering themselves as members of groups, Swedish citizens have begun to identify themselves as individuals and concerned citizens who can exercise political influence in new ways. A key concept in the emerging new collective identity is individual responsibility, community action, and civic republicanism. The non-socialist government encouraged this development by stressing the importance of the volunteering sector and proclaiming that civil society will be larger than government in non-socialist Sweden.

Social democratic political hegemony has been made, maintained, and broken in the periods of civil society history discussed in this book. Important actors in its making are the collective action organizations that emerged in the first, second, and third waves of organizational development. Important events include the struggle for universal suffrage, wartime crises, and construction of the welfare state. Actors of special importance are the Social Democratic Party and LO. Collective affiliation and the large number of collective action organizations that identified with the labor movement gave the social democrats a support mobilization unequaled by any other party or movement. Other organizations of civil society played a supportive role in the making and maintaining of social democratic political hegemony. Particular mention should be made of the white-collar union movement and even segments of organized business.

Actors of civil society have participated in breaking social democratic political hegemony. Many fourth wave organizations are important here. The new left and right have mobilized support for interpretations of Swedish history that differ considerably from the one offered by social democracy. Class solidarity is gradually being replaced by a multitude of new identities based on gender, community, ethnicity, and age. Organized business– and particularly the SAF– deserve special mention in the struggle over Swedish political hegemony. Established political actors have been involved in hegemonic struggles on various fronts, even including such social democratic strongholds as unions and workplaces. Both the Conservative and Center Parties have tried to mobilize support among working class voters, and the Liberal Party was the first non-socialist party to criticize corporatism directly. Other battlegrounds are the op-ed pages of newspapers, parliament, and election campaigns. Hegemonic issues have included the role of public sector, regulatory policies, labor laws, and corporatism.

The labor movement has inadvertently made its own contributions to the erosion of social democratic political hegemony through public scandals and, together with its inability to renew itself, this has lost it support. Among the problems are difficulty in demarcating the role of the party and union movement in politics. There is considerable talk but few concrete proposals on a clearer division of responsibility here. The LO still relies on the party even though the political opportunity structure of Sweden has changed dramatically. It maintains ties with the party though it is forced by its membership to focus more directly on narrower union matters. The decision to disband the system of collective affiliation between the LO and party meant that 800,000 party members (two-thirds of the party) have not renewed their membership. The question is how long the LO can remain ideologically aligned to the party and still reflect the concerns of its membership. Political activity in and identification with the Social Democratic Party has, as discussed earlier, been stronger in the high echelons of the unions and correspondingly much lower at the grassroot's level.

First dimension politics in the sixth period have become a struggle over political hegemony. Partisan politics in the early 1990s was a tug of war over political prestige. Pragmatic politics did not characterize Sweden in these years. Politicians ceased to function as rational actors which 'follow the principle of letting bygones be bygones, cutting one's losses and ignoring sunk costs'. Instead, they became hegemony fighters, 'relentless pursuer[s] of revenge', and hegemonic maintainers who 'would rather fight and lose than not fight at all' (Elster, 1989, p. 99). Emotion replaced common sense in Swedish politics. No wonder several collective action organizations began to reevaluate their relationships with government and political parties and newer ones disrespect national and local politicians. It is too early to evaluate whether this trend has been broken by the new social democratic government.

The history of civil society shows that struggles over political hegemony occur in times of great transformation. They entail definitions of collective identity, social norms, and political correctness. The struggle concerns the cement of society (Elster, 1989, p. 102). Dominant hegemonic actors formulate these rules, but their ability to maintain them and regulate behavior diminishes once individual and group preferences undergo transformation. Transformation here may come about through scientific, technical, economic, and social development. It concerns changes in material condition and the emergence of public issues that the existing political order is ill-designed to solve. Such developments tend to erode the ability of political hegemons to make credible threats and promises.

Developments in Sweden as well as elsewhere are showing that political institutions and citizens are beginning to play a different role in society, economics, and politics. Government is now less engaged in governance, i.e., the exercise of political responsibility for national well-being, and it seems that this trend will continue. Other actors are accepting responsibility for solving political, economic, and social problems. The 'skill revolution' which is empowering citizens through education to know when, where, and how to engage in collective action, limits to public growth, and the inability of political institutions to renew themselves to take responsibility for new social, political, and economic problems are furthering this development (cf. Roseanau, 1992). Problems of ungovernability have created an explosive vacuum which is now being filled by newer forms of collective action. This situation is reminiscent of the first two periods of civil society history.

Changes such as these indicate that the future political order will not be dominated by a distinctive, hegemonic power relationship with a single dominant actor. Research has shown that the labor movement has lost its innovative capacity, and class interests are no longer wedded to national interests. No single force has emerged to replace it, and it is unlikely that one single force will. Collective action organizations must adapt to the new political opportunity structure by using the choice opportunities for political access and action which are becoming available to them. Key characteristics for success are independence from government, ideological and organizational flexibility, problem orientation, access to all political parties, opinion formation, and dedication to the role of individual members to help solve the problems threatening Swedish welfare and culture.

Historical forgetting and glorification of social movements

The late 1800s were the most important decades for the formation of Swedish civil society. All the social movements that formed at this time were moving in the same direction and for the same general reasons. The main goal behind organizing efforts, demonstrations, and strikes was the struggle for freedom and respect; the enemy was privilege, despotism, and arbitrariness. The ideas of these first social movements fell on fertile soil. All movements filled a societal need, and they encouraged common women and men to assume responsibility for themselves and demand changes in the political order. They were oriented toward *revival,* i.e., a general awakening of Swedes to their rights, responsibilities, and potentials. Today this viewpoint is called civil republicanism. The movements were effective cultural creators. No wonder books on this period are given such imposing titles as 'A Changed World' (Gårdlund, 1945), 'They Showed the Way' (Svensson, 1979), and 'We Create Our Country' (Hirdman, 1988) (my translations). These movements were the vanguard of Sweden and represented the first and second waves of organizational development. They created a life style for their members and gave the new middle class– teachers, foremen, manufacturers, directors, and shopkeepers– and new working class– industrial workers– of the periods a setting to express their needs and realize their expectations. Each movement created its own language and architecture, and Swedes were instilled by their common values. First and second wave movements have been extremely conscious of their role in the early years of civil society history. Several writers have been commissioned by them to author historical accounts of the organizations and their leaders.

Members of the first social movements were the bearers of the new political ideologies– i.e., liberalism and socialism– that thrived as Sweden became an industrialized welfare state. International trends played an important role here. Also, demographic changes caused by industrialization led to feelings of rootlessness among people who relocated in the new settings of company towns and urban areas. They sought contact with other people with similar thoughts, feelings, and experiences. This need for a sense of community provided the conditions for the new ideas upon which the new collective action organizations of the time were founded. Tage Erlander, one of the most important Swedish political leaders of this century, captured the significance of the first social movements in the following words:

> For many people from the provence of Värmland, the large social movements of the 1800s– the free churches and teetotaling societies– appeared to be a force that saved family, friends, and the entire village from decline and destruction. These social movements taught cooperation and coordination. Poor people united at the mission houses and teetotaling chapters (Erlander, 1977, p. 31, my translation).

Collective action led to marked change in the character of government in many periods of civil society history analyzed in this book. Issues once considered to be outside the realm of politics became central questions for government. Over the years, the efforts by collective action organizations led to government expansion and development of the welfare state. Later welfare tasks were transferred from collective action organizations to government authorities.

The first and second periods of civil society history were the first golden years of Swedish collective action. The emergence and role of collective action organizations in politics, society, and economics in these years explains the proclivity by Swedes to extol them. Their successes in broadening the boundary of the political are considered a mark of national achievement and pride. Success here is seen as the groundwork of the people's home and prerequisites for the development of the Swedish welfare state. Swedes boast about the struggles for recognition which were fought by these early collective action organizations. They speak with pride about the commitment of citizens to these movements. Individual leaders are singled out for special recognition, and they are remembered in statues centrally placed in Swedish cities, streets named after them, and in biographies over their ideas and deeds.

Yet emphasis is seldom given to the problems which movement organizations later experienced. Neither is mention made of the negative effects of the Swedish model on collective action organization development. Stagnation of the first and most renowned social movements is a subject slighted in historical accounts. Another topic which only recently has been given attention is the now conservative and traditionalist nature of glorified collective action organizations. Today they are allied with the political order and have, because of this, lost most of their social movement character. The first wave movements as well as other organizations are highly reliant on government funds, as discussed in a earlier chapter section. Moreover, most of these movement organizations are very critical to new collective action organizations, which fill a function in contemporary Swedish society very similar to one once played by the now established and glorified movements. Problems with member recruitment, discrimination of women, neglect of the youth, and disregard of immigrants have made the established movements less a school for democracy and citizen empowerment of the less fortunate and more an institution for successful citizens. For instance, well-educated people dominate the courses given by adult education societies and sports associations for young people cater more to middle than working class children.

General tendencies toward defensive organizational maintenance and reliance on government patronage help explain the neglect of these important issues and problems by collective action organizations. But why have scholars ignored them? The answer is a complex one.

First, many Swedish scholars who have written on collective action organizations– and particularly social movements– have sympathized with them and not researched them in a scientific way. They have contributed to their glorification of these movements. Second, with few exception, scholars of Swedish collective action organizations have been historians who have focused research attention on the first golden years of Swedish civil society. A third reason is the lack of theoretical knowledge on the part of organizational scholars. Academics have not developed theoretical perspectives to analyze collective action organizations. No scholar has previously in any comprehensive way analyzed the history of Swedish civil society in terms of the four interdependent relationships of importance for this book: (1) the relationship between members and leaders, (2) the relationship among collective action organizations and between organizations and political parties; (3) the relationship between organizations and government, and (4) the relationship that organizations have with the mass media and general public. Neither have scholars analyzed civil society in terms of organizational waves of development, political opportunity structures, and the life cycle model.

Swedish scholars have generally assumed that collective action organizations are functioning properly and performing valuable political, economic, and social tasks. They have also defended corporatism and social democratic rule without much reflection. Unfortunately, there are scholars who, without much scientific basis for their conclusions, praise the now old and condemn contemporary new collective action organizations. Their efforts have not furthered scholarship in this field. Of course, there are notable exceptions here and their work has been cited in this book, but the general tendency has been for scholars to contribute to historical forgetting by closing their minds– and therefore ours– to developments contrary to their preconceptions. They have romanticized collective action organizations. This book and the companion volume *Understanding Organized Action* attempt to rectify these errors, and it is hoped that scholars who investigate the now new collective action organizations will use the knowledge produced in them to avoid the mistakes of their predecessors.

Appendix 1 Membership rates of selected free churches

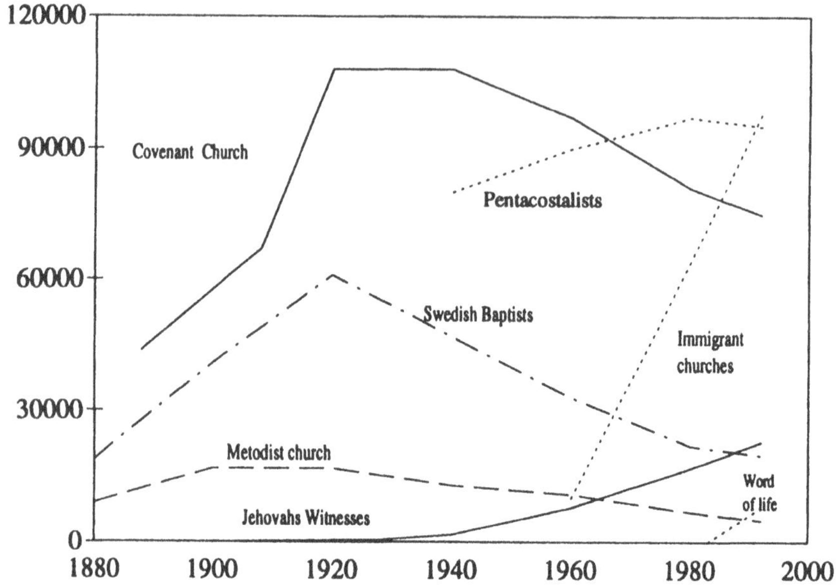

Source: Statistisk årsbok (SCB), Religionssociologiska institutet

Appendix 2 Membership rates of selected teetotaling organizations

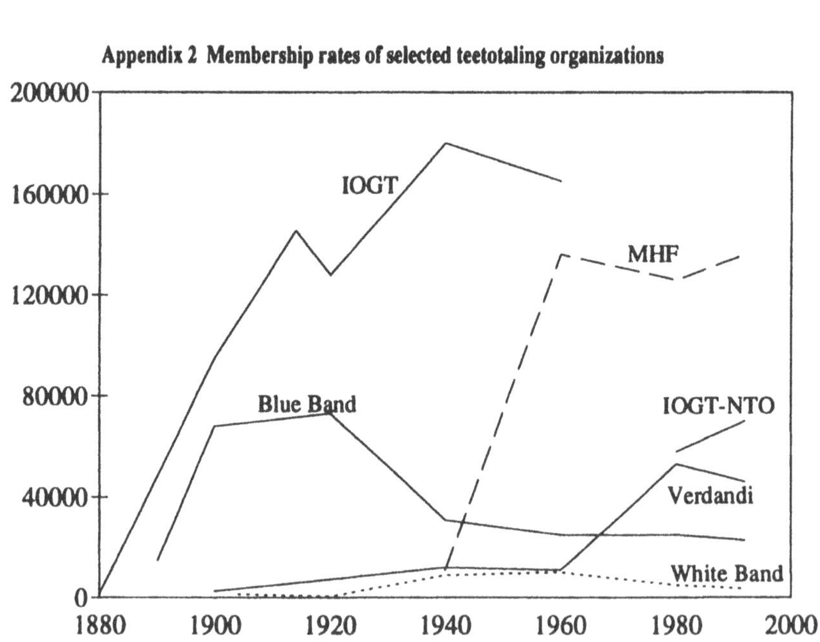

Source: Statistisk årsbok (SCB), statistics from the organizations

Appendix 3 Membership rates of peak labor unions

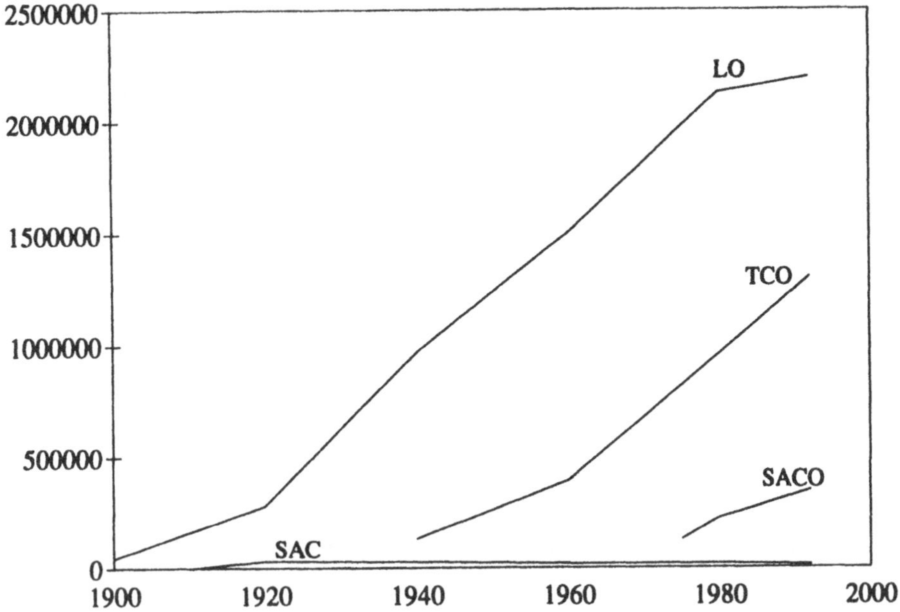

Source: Statistics from the organizations

Appendix 4 Membership rates of selected women's organizations

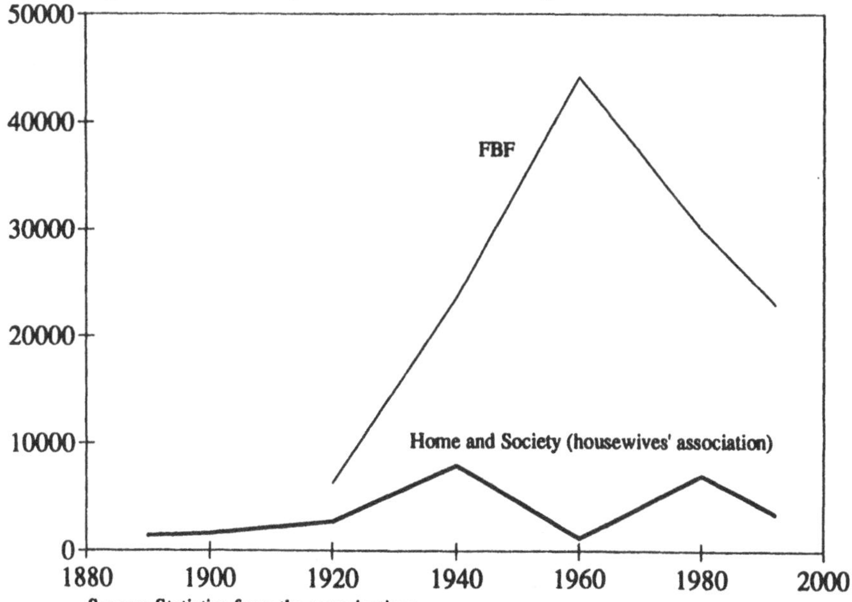

Source: Statistics from the organizations

Appendix 5 Membership rates of selected peace organizations

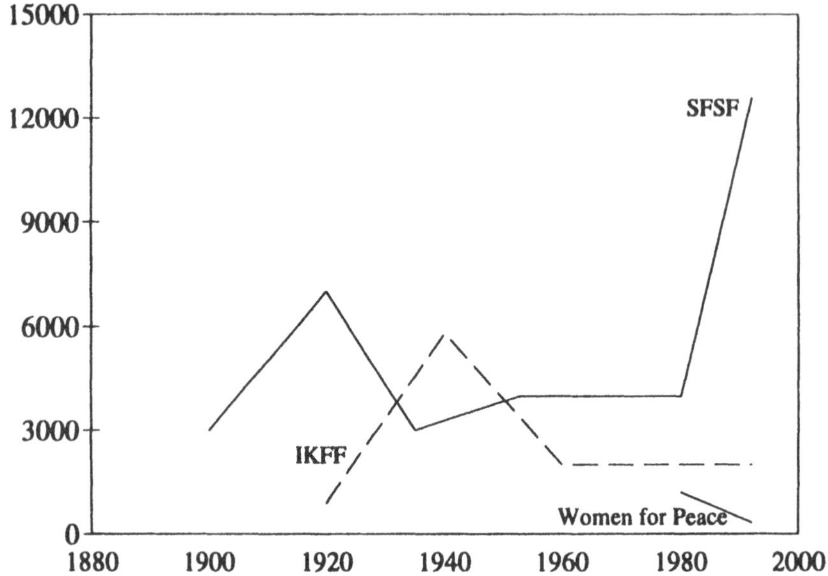

Source: Statistics from the organizations

Appendix 6 Membership rates of selected consumer and housing cooperatives

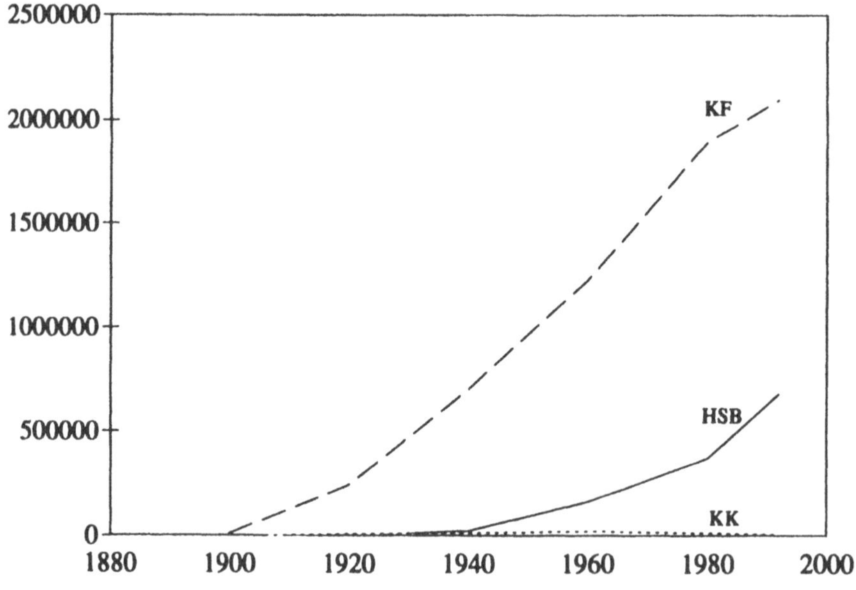

Source: Statistics from the organizations, Statistisk årsbok (SCB)

Appendix 7 Membership rates of sports' movement

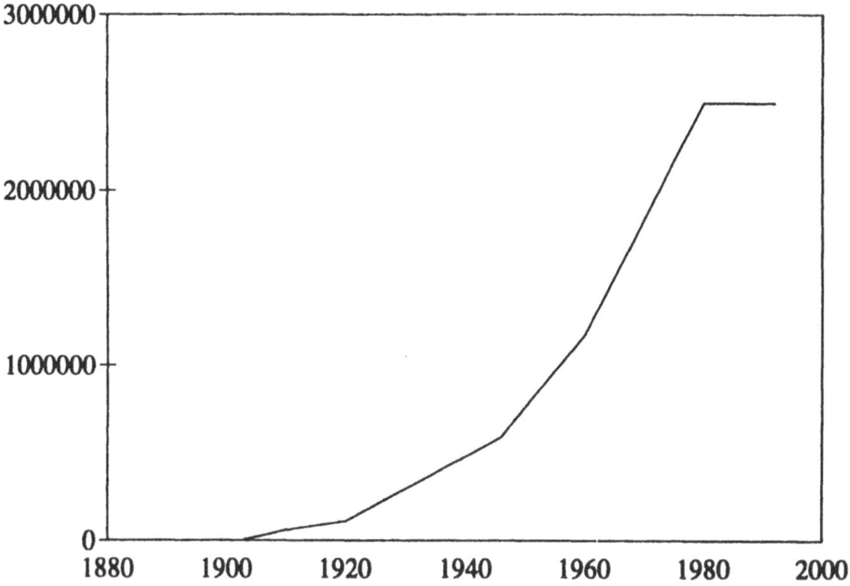

Source: Statistics from the organization

Appendix 8 Membership rates of selected environmental organizations

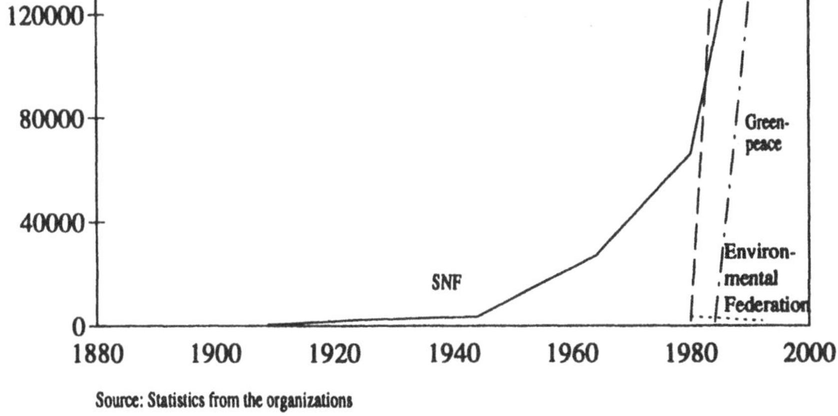

Source: Statistics from the organizations

Appendix 9 Membership rates of selected humanitarian organizations

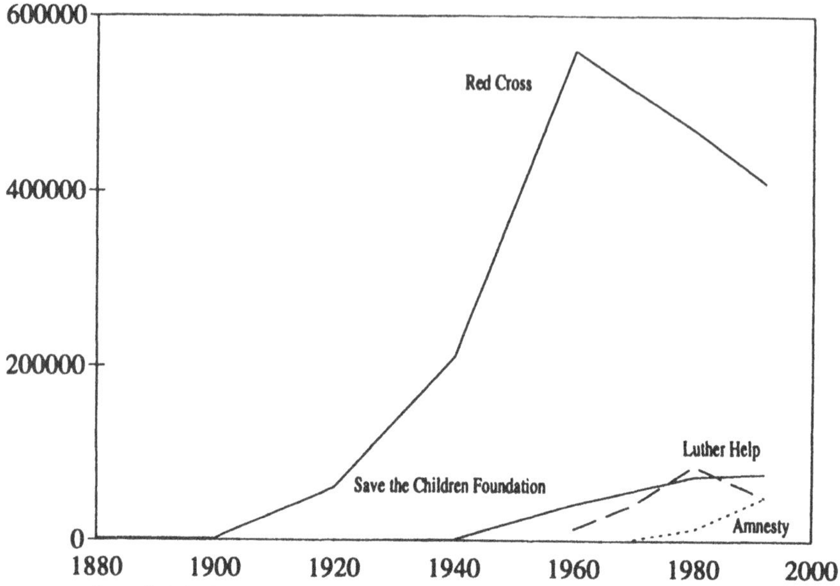

Source: Statistics from the organizations

Appendix 10 Membership rates of selected associations for senior citizens

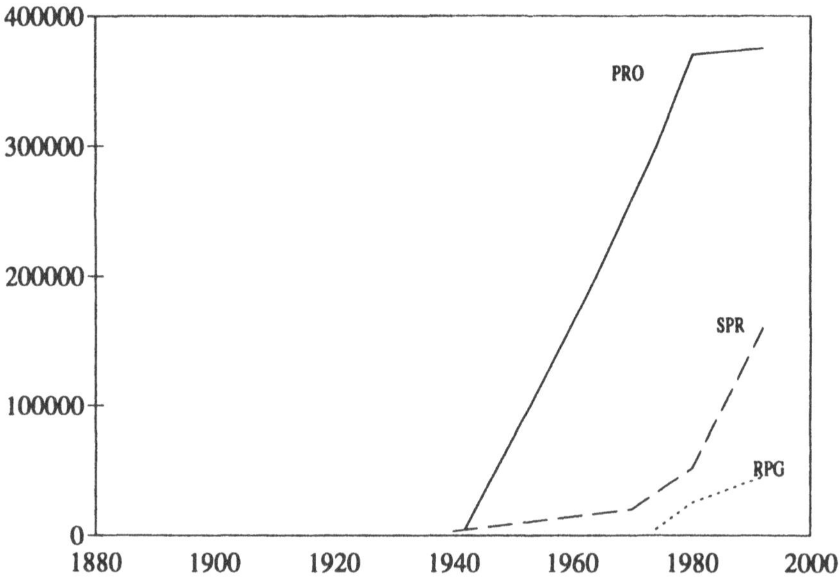

Source: Statistics from the organizations

References

Ahlén, Kristina (1989), 'Swedish Collective Bargaining Under Pressure: Inter-union Rivalry and Incomes Policies', *British Journal of Industrial Relations* vol. 27: 330-46.

Ahlén, Kristina (1990), *Changing the Value of Union Membership with Marketplace Benefits*, Center of Working Life, 'Unions Facing Change Project', Report No. 1, Stockholm.

Ahrne, Göran (1987), 'Byråkrati i organisationer och i samhället' in Ulla Bergryd, ed.

Åkerman, Brita (1944), 'Genom kunskap till medborgarskap' in Margareta von Konow, ed.

Allardt, Erik (1977), *Att ha, att älska, att vara. Om välfärd i Norden*, Argos, Lund.

Åmark, Klas (1986), *Facklig makt och fackligt medlemskap. De svenska fackförbundens medlemsutveckling 1890-1940*, Arkiv, Lund.

Anners, Erik (1966), 'Förord' in Jan-Olof Sundell.

Anton, Thomas J. (1969), 'Policy-Making and Political Culture in Sweden', *Scandinavian Political Studies* vol. 4: 88-102.

Arbetet (1963), "'Utspelad' opposition", 15 February.

Arvedson, Lennart, Ingemund Hägg, and Bengt Rydén, eds. (1983), *Land i olag. Samhällsorganisation under omprövning*, SNS: Stockholm.

Åström, Gertrud and Yvonne Hirdman, eds. (1992), *Kontrakt i kris. Om kvinnors plats i välfärdsstaten*, Carlsson, Stockholm.

Bäck, Henry (1983), *Invandrarnas riksorganisationer*, Delegationen för invandrarforskning, Stockholm.

Barnen & Vi (Save the Children Foundation) (1990), 'Rädda barnens resultat 1989', no. 4.

Beckford, James A. (1985), *Cult Controversies. The Societal Response to New Religious Movements*, Tavistock Publications, London.

Bennulf, Martin (1990a), 'Det gröna genombrottet' in Mikael Gilljam and Sören Holmberg, eds.

Bennulf, Martin (1990b), 'Förtroende för Greenpeace' in Sören Holmberg and Lennart Weibull, eds.

Bergqvist, Christina (1991), 'Corporatism and Gender Equality. A Comparative Study of Two Swedish Labour Market Organizations', *European Journal of Political Research* vol. 20: 107-25.

Bergqvist, Sven-Runo (1969), *Kommunerna och organisationerna*, University of Uppsala, Department of Political Science, Unpublished masters' thesis, Uppsala.

Bergryd, Ulla, ed. (1987), *Den sociologiska fantasin– teorier om samhället*, Rabén & Sjögren, Stockholm.

Birkinshaw, Patrick (1988), 'Corporatism and Accountability' in Andrew Cox and Noel O'Sullivan, eds.

Bjereld, Ulf and Mikael Gilljam (1991), 'Vad vill de kristna väljarna? Religionens betydelse för människors åsikter i politiska sakfrågor', *Statsvetenskaplig tidskrift*: 149-59.

Blomdahl, Ulf (1990), *Folkrörelserna och folket– med utblick mot framtiden*, Carlsson, Stockholm.

Bodegård, Anders and Styrbjörn Gustavsson (1974), 'LO, ABF och de utländska arbetarna', *Häften för kritiska studier* no. 2: 33-47.

Boglind, Anders (1989), *Medarbetare och medlem. Individuell och kollektiv rationalitet i arbetslivet*, University of Gothenburg, Department of Sociology, Gothenburg.

Bosworth, Barry P. and Alice M. Rivlin, eds. (1987), *The Swedish Economy*, The Brookings Institution, Washington, D. C.

Brand, Karl-Werner (1990), 'Cyclical Aspects of New Social Movements: Waves of Cultural Criticism and Mobilization Cycles of New Middle-Class Radicalism' in Russel J. Dalton and Manfred Keuchler, eds.

Broström, Anders, ed. (1981), *Storkonflikten 1980 på den svenska arbetsmarknaden*, Arbetslivscentrum, Stockholm.

Carle, Jan and Per Sjöstrand (1991), *Ungdom, politik, inflytande*, Statens Ungdomsråd, Stockholm.

Cartwright, Dorwin and Alvin Zander (1968), *Group Dynamics: Research and Theory*, Harper & Row, New York.

Casparsson, Ragnar (1966), *Saltsjöbadsavtalet i historisk belysning*, Tiden, Stockholm.

Child, Marquis (1937), *Sweden: The Middle Way*, Yale University Press, New Haven.

Child, Marquis (1980), *Sweden: The Middle Way on Trial*, Yale University Press, New Haven.

Coleman, Simon (1991), 'Livets Ord och det svenska samhället. Analys av en debatt', *Tro & Tanke* no. 4: 31-73.

Cox, Andrew and Noel O'Sullivan, eds. (1988), *The Corporate State. Corporatism and the State Tradition in Western Europe*, Edward Edgar, Aldershot.
Czanda, Roland M. and Andrienne Windhoff-Héritier, eds. (1991), *Political Choice. Institutions, Rules, and the Limits of Rationality*, Campus Verlag, Frankfurt am Main.
Dahlström, Edmund et al. (1971), *LKAB och demokratin. Rapport om en strejk och ett forskningsprojekt*, W & W-serie, Stockholm.
Dalton, Russell J. and Manfred Kuechler, eds. (1990), *Challenging the Political Order. New Social and Political Movements in Western Democracies*, Polity Press, Oxford.
De Geer, Hans (1989), *I vänstervind och högervåg. SAF under 1970-talet*, SAF, Stockholm.
De Geer, Hans (1991), 'Corporatism and Neutrality: Sweden during the Second World War' in Wyn Grant, Jan Nekkers, and Frans Van Waarden, eds.
Den svenska historien. Del 9. Industri och folkrörelser 1866-1920, Bonniers, Stockholm.
Edlund, Sten et al. (1989), *Saltsjöbadsavtalet 50 år. Forskare och parter begrundar en epok 1938-1988*, Arbetslivscentrum, Stockholm.
Eduards, Maud (1992), 'Den feministiska utmaningen: Kvinnors kollektiva handlande' in Gertrud Åström and Yvonne Hirdman, eds.
Elliot, Maria (1992), 'Förtroende på fallrepet' in Sören Holmberg and Lennart Weibull, eds.
Elmbrant, Björn (1964), *Antikärnvapenrörelsen i Sverige. Ett försök till tvärsnitt*, University of Stockholm, Department of Political Science, unpublished undergraduate paper, Stockholm.
Elster, Jon (1989), *The Cement of Society. A Study of Social Order*, Cambridge University Press, Cambridge.
Elvander, Nils (1972), *Intresseorganisationerna i dagens Sverige*, Gleerups, 2nd edition, Lund.
Elvander, Nils (1988), *Den svenska modellen. Löneförhandlingar och inkomstpolitik 1982-1986*, Allmänna Förlaget, Stockholm.
Engellau, Patrik and Ulf Henning, eds. (1984), *Nordic Views and Values*, Nordic Council, Stockholm.
Engberg, Jan et al. (1978), *Utanför systemet. Vänstern i Sverige 1968-78*, Rabén & Sjögren, Stockholm.
Engberg, Jan (1985), 'Den privata internationalismen: Organisationssverige söker sig utomlands', *Statsvetenskaplig tidskrift*: 341-8.
Eriksen, Sidel (1990), 'Drunken Danes and Sober Swedes? Religious Revivalism and the Temperance Movement as Keys to Danish and Swedish Folk Culture' in Bo Stråth, ed.

Eriksson, Björn et al. (1982), *Det förlorade försprånget*, Miljöförbundet, Gothenburg.
Eriksson, Björn (1982), 'Det var så det började' in Björn Eriksson et al.
Eriksson, Sune (1962), 'Den sociala hemhjälpen' in Svenska Landskommunernas Förbund.
Erlander, Tage (1977), *1901-1939*, Tiden, Stockholm.
Esaiasson, Peter (1990), *Svenska valkampanjer 1866-1988*, Allmänna Förlaget, Stockholm.
Feldt, Kjell-Olof (1991), *Alla dessa dagar... I regeringen 1982-1990*, Norstedts, Stockholm.
Fi (1992), Written Statements on Promemoria om intresserepresentation i vissa verksstyrelser mm (Dnr 1062/92), Stockholm, Ministry of Finance.
Fi Ds 1992, no. 114, *Statens bidrag till kvinnoorganisationernas centrala verksamhet*, Ministry of Finance, Stockholm.
Folkstyrelsekommittén (1985), *Makten från folket. Tolv uppsatser om folkstyrelsen*, Liber, Stockholm.
Förslag till lagar om registrerade föreningar för anna än ekonomisk verksamhet och om offentligen erkända stiftelser (1903), Swedish Parliament, Stockholm.
Foyer, Lars (1961), 'Former för kontakt och samverkan mellan staten och organisationerna' in SOU, 1961, no. 21.
FR (1958), *Frikyrkan inför framtiden*, Stockholm: FR.
Frängsmyr, Tore, Ingemund Hägg, and Bengt Rydén, eds. (1984), *Kan Sverige styras? Att förstå och styra samhället*, SNS, Stockholm.
Frikyrkomötet (1964), *Föredrag och protokoll från det tionde frikyrkomötet i Immanuelskyrkan, Stockholm, den 10-12 mars 1964*, Frikyrkomötet, Stockholm.
FS (1970), *Vad vill frikyrkan? Kyrka. Stat. Frågan*, FS, Stockholm.
Gamson, William A. and Andre Modigliani (1971), *Untangling the Cold War. A Strategy for Testing Rival Theories*, Little, Brown and Co., Boston.
Gårdlund, Torsten (1942), *Industrialismens samhälle*, Tiden, Stockholm.
Gårdlund, Torsten (1945), *En förändrad värld*, Kooperativa Förbundet, Stockholm.
Gardner, James N. (1991), *Effective Lobbying in the European Community*, Kluwer, Boston.
Gelb, Joyce (1987), "Social Movement 'Success': A Comparative Analysis of Feminism in the United States and the United Kingdom" in Mary Fainsod Katzenstein and Carol McClurg Mueller, eds.
George, Stephen (1985), *Politics and Policy in the European Community*, Clarendon Press, Oxford.
Giddens, Anthony (1990), *The Consequences of Modernity*, Polity Press, Oxford.
Gilljam, Mikael (1988), *Svenska folket och löntagarfonderna. En studie i politisk åsiktsbildning*, Studentlitteratur, Lund.

Gilljam, Mikael and Sören Holmberg (1990), *Rött, blått, grönt. En bok om 1988 års riksdagsval*, Bonniers, Stockholm.
Gilljam, Mikael and Sören Holmberg (1993), *Väljarna inför 90-talet*, Norstedts Juridik, Stockholm.
Grant, Wyn, ed. (1985), *Political Economy of Corporatism*, Macmillan, London.
Grant, Wyn, Jan Nekkers, and Frans Van Waarden, eds. (1991), *Organising Business for War. Corporatist Economic Organization during the Second World War*, Berg Publishers, New York.
Greenwood, Justin, Jürgen R. Grote, and Karsten Ronit, eds. (1992), *Organized Interests and the European Community*, Sage, London.
Gustafsson, Hans (1977). Speech, Stockholm, Ministry of Municipalities, RGSM, no. 6, p. 4.
Haas, Ernst B. (1968, 1958), *The Uniting of Europe. Political, Social, and Economic Forces 1950-1957*, Stanford University Press, Stanford.
Hall, Johan (1992), *LO och löntagarfonder*, University of Stockholm, Department of Political Science, Report from the project 'Swedish Interest Organizations', Stockholm.
Hallenstvedt, Abraham (1974), *Nordisk foreningsliv*, Institutt for samfunnsforskning, Oslo.
Hansen, J. A., ed. (1992), *Environmental Concerns. An Inter-disciplinary Exercise*, Elsevier Applied Science, London.
Hansson, Sigfrid (1936), 'Fackföreningsrörelsen' in *Svenska folkrörelser 1*.
Haraldsson, Désirée (1987), *Skydda vår natur! Svenska Naturskyddsföreningens framväxt och tidiga utveckling*, Lund University Press, Lund.
Heckscher, Gunnar (1951), *Staten och organisationerna*, Kooperativa förbundets bokförlag, Stockholm.
Heller, Agnes and Ferenc Fehér (1988), *The Postmodern Political Condition*, Polity Press, London.
Hernes, Helga Maria (1987), *Welfare State and Woman Power. Essays in State Feminism*, Norwegian University Press, Oslo.
Hirdman, Yvonne (1988), *Vi bygger landet. Den svenska arbetarrörelsens historia från Per Götrek till Olof Palme*, Tiden, Stockholm.
Holm, Olof (1989), Organisation i dissonans– konsumentrörelse i brytningstid, Timbro, Stockholm.
Holmberg, Sören and Kent Asp (1984), *Kampen om kärnkraften. En bok om väljare, massmedier och folkomröstningen 1980*, Publica, Stockholm.
Holmberg, Sören och Mikael Gilljam (1987), *Väljare och val i Sverige*, Bonniers, Stockholm.
Holmberg, Sören and Peter Esaiasson (1988), *De folkvalda. En bok om riksdagsledamöterna och den representative demokratin i Sverige*, Bonniers, Stockholm.

Holmberg, Sören and Lennart Weibull, eds. (1990), *Medier och opinion i Sverige*, University of Gothenburg, Department of Political Science, Gothenburg.
Holmberg, Sören and Lennart Weibull, eds. (1992), *Trendbrott*, University of Gothenburg, Department of Political Science, Gothenburg.
Huntford, Roland (1971), *The New Totalitarians*, Allen Lane, London.
Hurd, Madeleine (1990), 'A Tale of Two Cities: Socialists, Liberals and Democracy in Hamburg and Stockholm, 1870-1914' in Bo Stråth, ed.
IF (1980), Letter from Bo Helmerson (head of the IF energy and environment unit) to the General Secretary of the National Committee for Alternative 2 in the Nuclear Power Referendum, 5 February.
Isberg, Magnus (1982), *The First Decade of the Unicameral Riksdag. The Role of the Swedish Parliament in the 1970s*, University of Stockholm, Department of Political Science, Report no. 1, Stockholm.
Jamison, Andrew, Ron Eyerman, Jacqueline Cramer with Jeppe Læsse (1990), *The Making of the New Environmental Consciousness. A Comparative Study of the Environmental Movements in Sweden, Denmark and the Netherlands*, Edinburgh University Press, Edinburgh.
Johansson, Alf (1962), 'Bostadspolitiken' in Svenska Landskommunernas Förbund.
Johansson, Alf (1989), 'Arbetarklassen och Saltsjöbaden 1938: perspektiv ovan- och underifrån' in Sten Edlund et al.
Johansson, Anders L. (1989a), *Tillväxt och klassamarbete - en studie av den svenska modellens uppkomst*, Tiden, Stockholm.
Johansson, Anders L. (1989b), 'Saltsjöbadspolitikens förhistoria' in Sten Edlund et al.
Johansson, Hilding (1947), *Den svenska godtemplarrörelsen och samhället*, Oskar Eklunds Bokförlag, Stockholm.
Johansson, Jan (1992), *Det statliga kommittéväsendet. Kunskap, kontroll, konsensus*, University of Stockholm, Department of Political Science, Stockholm.
Johansson, Inge (1992), 'Den mödosamma vägen tillbaka', *Fönstret* (ABF's newspaper) no. 1: 9-10.
Jones, Kathleen B. and Anna G. Jónasdóttir, eds. (1988), *The Political Interests of Gender. Developing Theory and Research with a Feminist Face*, Sage, London.
Källstad, Thorvald (1964), 'Frikyrkornas samhällsansvar' in Frikyrkomötet.
Karvonen, Lauri (1981), *'Med vårt västra grannland som förebild' En undersökning av policydiffusion från Sverige till Finland*, Åbo Akademi, Åbo.
Katzenstein, Mary Fainsod (1987), 'Comparing the Feminist Movements of the United States and Western Europe. An Overview' in Mary Fainsod Katzenstein and Carol McClurg Mueller, eds.
Katzenstein, Mary Fainsod and Carol McClurg Mueller, eds. (1987), *The Women's Movements of the United States and Western Europe. Consciousness,*

Political Opportunity, and Public Policy, Temple University Press, Philadelphia.

Katzenstein, Peter J. (1985), *Small States in World Markets. Industrial Policy in Europe*, Cornell University Press, Ithaca.

Klandermans, Bert, Hanspeter Kriesi, and Signey Tarrow (1988), *International Social Movement Research Vol. 1. From Structure to Action: Comparing Social Movement Research Across Cultures*, JAI Press, Greenwich.

Koblic, Steven, ed. (1975), *Sweden's Development From Poverty to Affluence, 1750-1970*, University of Minnesota Press, Minneapolis.

Kodama, Katsuya and Unto Vesa, eds. (1990), *Towards a Comparative Analysis of Peace Movements*, Gower Publishing Company, Brookfield.

von Konow, Margareta (1944), 'Förord' in Margareta von Konow, ed.

von Konow, Margareta, ed. (1944), *I Fredrika Bremers spår. Fredrika-Bremer-Förbundet 1884-1944*, FBF, Stockholm.

Konsum (1988), *Medlemsdemokratiska kommitténs slutrapport*, Konsum, Stockholm.

Korpi, Walter (1974), *Varför strejkar arbetarna? En sociologisk betraktelse över konflikter, makt och rätt på arbetsmarknaden*, Tiden, Stockholm.

Korpi, Walter (1978), *Arbetarklassen i välfärdskapitalismen. Arbete, fackförening och politik i Sverige*, Prisma, Stockholm.

Kron, Fredrik (1992), *Svenska freds- och skiljedomsföreningen– en studie*, University of Stockholm, Department of Political Science, Report from the project 'Swedish Interest Organizations', Stockholm.

Lane, Jan-Erik, ed. (1991), *Understanding the Swedish Model*, Frank Cass, London.

Larsson, Janerik (1983), *Vändpunkten. Boken om 4 oktober 1983*, Timbro, Stockholm.

Lewin, Leif (1967), *Planhushållningsdebatten*, Almqvist & Wiksell, Stockholm.

Lewin, Leif (1975), 'The Debate on Economic Planning in Sweden' in Steven Koblic, ed.

Lijphart, Arend and Markus M. L. Crepaz (1991), 'Corporatism and Consensus Democracy in Eighteen Countries: Conceptual and Empirical Linkages', *British Journal of Political Science* vol. 21: 235-56.

Lindahl, Rutger (1992), 'När opinionen vände: Åsikter i EG-frågan', in Sören Holmberg and Lennart Weibull, eds.

Lindblad, Ingemar (1960), *Svenska kommunalarbetarförbundet 1910-1960. En studie i svensk fackföreningsrörelse*, Tiden, Stockholm.

Lindfeldt, Erik (1979), 'Från herbarium till miljökamp - om fältbiologerna i miljörörelsen', *Natur och samhälle* no. 3: 124-39.

Lindkvist, Kent (1990), 'Mobilization Peaks and Declines of the Swedish Peace Movement' in Katsuya Kodama, ed.

Lindorm, Erik (1958), *Ny svensk historia. Ett folk på marsch 1914-1932*, Bonniers, Stockholm.
Lindorm, Erik (1961), *Ett folk på marsch 1810-1872*, Bonniers, Stockholm.
Lindroth, Jan (1974), *Idrottens väg till folkrörelse. Studier i svensk idrottsrörelse till 1915*, University of Uppsala, Department of History, Uppsala.
Lipset, Seymour Martin, ed. (1986), *Unions in Transition. Entering the Second Century*, ISC Press, San Francisco.
LO (1989), *Rapport från arbetsgruppen PUST*, LO, Stockholm.
Logue, John (1980), *Toward a Theory of Trade Union Internationalism*, University of Gothenburg, Department of History, Gothenburg.
Lundkvist, Sven (1968), 'Nykterhetsrörelsen' in *Den svenska historien. Del 9.*
Lundkvist, Sven (1977), *Folkrörelserna i det svenska samhället 1850-1920*, Almqvist & Wiksell International, Stockholm.
Marcusson, Lena and Olof Petersson (1990), 'Orienteringspunkter i den grå zonen', *Nordisk Administrativt Tidskrift* vol. 3: 288-95.
McEachern, Doug (1990), *The Expanding State. Class and Economy in Europe Since 1945*, Harvester Weatsheaf, New York.
Meijer, Hans (1956), *Kommittépolitik och kommittéarbete. Det statliga kommittéväsendets utvecklingslinjer 1905-1954 samt nuvarande funktion och arbetsformer*, C. W. K. Gleerup, Lund.
Meijer, Hans (1969), 'Bureaucracy and Policy Formulation in Sweden', *Scandinavian Political Studies* vol. 4 (old series): 103-16.
Micheletti, Michele (1984a), 'The Involvement of Swedish Labor Market Organizations in the Swedish Political Process', SNS Occasional Paper No. 18, 1984.
Micheletti, Michele (1984b), *From Saltsjöbaden, to Harpsund, Haga, and the Fourth of October. Interest Organization and Politics in Sweden*, University of Stockholm, Department of Political Science, unpublished report, Stockholm.
Micheletti, Michele (1985a), 'Intresseorganisationerna och den allmänna opinionen' in Folkstyrelsekommittén.
Micheletti, Michele (1985b), *Organizing Interest and Organized Protest. Difficulties of Member Representation for the Swedish Central Organization of Salaried Employees (TCO)*, University of Stockholm, Sweden, Department of Political Science, Stockholm.
Micheletti, Michele (1987), 'Organization and Representation of Farmers' Interests in Sweden', *Sociologia Ruralis* vol. 27: 198-212.
Micheletti, Michele (1988a), 'De svenska intresseorganisationerna– i går, i dag och i morgon', *Statsvetenskaplig tidskrift* No. 1: 41-54.
Micheletti, Michele (1990a), *The Swedish Farmers' Movement and Government Agricultural Policy*, Praeger, New York.

Micheletti, Michele (1990b), 'Toward Interest Inarticulation: A Major Consequence of Corporatism for Interest Organizations', *Scandinavian Political Studies* vol. 13: 255-76.
Micheletti, Michele (1991a), 'Swedish Corporatism at a Crossroad: The Impact of New Politics and New Social Movements' in Jan-Erik Lane, ed.
Micheletti, Michele (1993), 'Sweden: Interest Groups in Transition and Crisis' in Clive S. Thomas, ed.
Micheletti, Michele (1994), *Understanding Organized Action*, University of Stockholm, Department of Political Science, Stockholm.
Misgeld, Klaus, Karl Molin, Klas Åmark, eds. (1992), *Creating Social Democracy. A Century of the Social Democratic Labor Party in Sweden*, Pennsylvania State University Press, University Park.
Möller, Yngve (1992), *Sverige och Vietnamkriget. Ett unikt kapitel i svensk utrikespolitik*, Tiden, Stockholm.
Mushaben, Joyce Marie (1983), 'The Forum New Dimensions of Youth Protest in Western Europe', *Journal of Political and Military Sociology* vol. 11: 123-44.
Närling, Ann (1989), *Är de starka männens tid förbi? En studie av fackligt ledarskap*, University of Stockholm, Department of Political Science, unpublished undergradute paper, Stockholm.
Närling, Ann (1993a), *Flytta på er gubbar!*, University of Stockholm, Department of Political Science, Report from the Project 'Swedish Interest Organizations', Stockholm.
Närling, Ann (1993b), *Artskillnader och gradskillnader i organisationernas opinionsbildning*, University of Stockholm, Department of Political Science, Report from the Project 'Swedish Interest Organizations', Stockholm.
Nilsson, Tommy (1985), *Från kamratföreningar till facklig rörelse. De svenska tjänstemännens organisationsutveckling 1900-1980*, Arkiv, Lund.
Norberg, Rolf och Lena (1966), *Pensionärernas Riksorganisation och partipolitik*, Forum för Borgerlig Debatt, Stockholm.
Nykterhetsrörelsens Landsförbund (1972), *Folkrörelserna och alkoholfrågan*, Nykterhetsrörelsens Landförbund, Stockholm.
Öberg, PerOla *(1991), Artikulering och kontroll av särintressen i svensk politik*, University of Uppsala, Department of Political Science, unpublished report, Uppsala.
Olsen, Johan P., ed (1990), *Demokrati på svenska*, Carlsson, Stockholm.
Olsen, Johan P., ed. (1991), *Svensk demokrati i förändring*, Carlsson, Stockholm.
Olson, Hans-Erik (1992), *Staten och ungdomens fritid. Kontroll eller autonomi?*, Arkiv, Lund.
Oskarsson, Maria (1990), 'Klassröstning på reträtt' in Mikael Gilljam and Sören Holmberg, eds.
Oskarsson, Maria (1992), 'Kvinnor– den offentliga sektorns försvarare' in Sören Holmberg and Lennart Weibull, eds.

Palm, Irving (1983), *Medlemsförändring genom flyttning. En frikyrkoförsamlings medlemsutveckling 1940-1980*, Religionssociologiska institutet, Report no. 177, Stockholm.
Panebianco, Angelo (1988), *Political Parties. Organization and Power*, Cambridge University Press, New York. Translated by Marc Silver.
Persson, Lennart K. (1975), *Syndikalismen i Sverige 1903-1922*, Federativs förlag, Stockholm.
Peterson, Abby (forthcoming), 'Women and the Social Movement Sector in Sweden', University of Gothenburg, Department of Sociology, unpublished report, Gothenburg.
Petersson, Olof (1979), *Regeringsbildningen 1978*, Rabén & Sjögren, Stockholm.
Petersson, Olof (1989), *Maktens nätverk. En undersökning av regeringskansliets kontakter*, Carlsson, Stockholm.
Petersson, Olof (1991a), *Nordisk politik*, Allmänna förlaget, Stockholm.
Petersson, Olof (1991b), 'Democracy and Power in Sweden', *Scandinavian Political Studies* vol. 14: 173-91.
Petersson, Olof, Anders Westholm, and Göran Blomberg (1989), *Medborgarnas makt*, Carlsson, Stockholm.
Pontusson, Jonas (1988), *Swedish Social Democracy and British Labour: Essays on the Nature and Conditions of Social Democratic Hegemony*, Cornell University, Center for International Studies, Western Societies Papers no. 19, Ithica.
Qvist, Gunnar (1974), *Statistik och politik. Landsorganisationen och kvinnorna på arbetsmarknaden*, Prisma and LO, Stockholm.
Reshef, Yonatan and Alan I. Murray (1988), 'Toward a Neoinstitutionalist Approach in Industrial Relations', *British Journal of Industrial Relations* vol. 26: 85-97.
Rosenau, James N. (1988), 'Patterned Chaos in Global Life: Structure and Process in the Two Worlds of World Politics', *International Political Science Review* vol. 9: 327-64.
Rosenau, James N. (1992), 'Governance, Order, and Change in World Politics' in James N. Rosenau and Ernst-Otto Czempiel, eds.
Rosenau, James N. and Ernst-Otto Czempiel (1992), *Governance Without Government: Order and Change in World Politics*, Cambridge University Press, Cambridge.
Rothstein, Bo (1986), *Den socialdemokratiska staten. Reformer och förvaltning inom svensk arbetsmarknads- och skolpolitik*, Arkiv, Lund.
Rothstein, Bo (1989), 'State and Capital in Sweden. The Importance of Corporatist Arrangements', *Scandinavian Political Studies* vol. 11: 235-60.
Rothstein, Bo (1992), *Den korporativa staten. Intresseorganisationer och statsförvaltning i svensk politik*, Norstedts, Stockholm.
Ruin, Olof (1983), 'Att komma överens och tänka efter före' in Lennart Arvedson, Ingemund Hägg, and Bengt Rydén, eds.

Ruth, Arne (1984), 'The Second New Nation: The Mythology of Modern Sweden', *Daedalus* vol. 113, No, 2: 53-96.

SAF (1991), *Färväl till korporatism*, SAF, Stockholm.

SAFAD (1991), *Statligt föreningsstöd– en kartläggning*, Report no. 6, SAFAD, Stockholm.

Sainsbury, Diane (1991), 'Swedish Social Democracy in Transititon: The Party's Record in the 1980s and the Challenge of the 1990s' in Jan-Erik Lane, ed.

Salomonson, Kurt [1991] (1958, 1959, 1962), *Mannen utanför, sveket, skiljevägen. Tre romaner om folkhemmet*, Norstedts, Stockholm.

Samhälle och riksdag. Del II, Almqvist & Wiksell, Stockholm.

Sandemose, Aksel (1984), 'The Law of Jante' in Patrik Engellau and Ulf Henning, eds.

Sargent, Jane A. (1985), 'Corporatism and the European Community', in Wyn Grant, ed.

SCB (1990), *Offentlig sektorn. Utveckling och nuläge*, SCB, Stockholm.

Scharpf, Fritz W. (1987), *Crisis and Choice in European Social Democracy*, Cornell University Press, Ithica.

Scharpf, Fritz W. (1991), 'Political Institutions, Decision Styles, and Policy Choices' in Roland M. Czanda and Adrienne Windhoff-Héritier, eds.

Schierup, Carl-Ulrik (1989), 'Immigrants and Immigrant Policy in Denmark and Sweden: Failed Assimilationism Versus Corporatist Multiculturalism', *International Review of Comparative Public Policy* vol. 1: 223-58.

Schiller, Bernt (1967), *Storstrejken 1909. Förhistoria och orsaker*, Akademiförlaget, Gothenburg.

Schiller, Bernt (1975), 'Years of Crisis, 1906-1914' in Steven Koblik, ed.

Schiller, Bernt (1981), 'Storstrejken 1909 och storkonflikten 1980. En jämförelse' in Anders Broström, ed.

Schiller, Bernt (1985), *Arbetare - Bönder - Företagare. En rapport om 'Interest Organizations and Internationalism'–ett samtidshistoriskt forskningsprojekt med förteckning av publikationer och material*, University of Gothenburg, Department of History, Report no. 9, Gothenburg.

Schiller, Bernt (1988), *'Det förödande 70-talet'. SAF och medbestämmandet 1965-1982*, Work Environment Fond, Stockholm.

Sidebäck, Göran (1992), *Kampen om barnets själ. Barn- och ungdomsorganisationer för fostran och normbildning 1850-1980*, Carlsson, Stockholm.

Siim, Birte (1988), 'Towards a Feminist Rethinking of the Welfare State' in Kathleen B. Jones and Anna G. Jónasdóttir, eds.

Sjöberg, Fredrik (1988), *På maktens tröskel. Miljörörelsen i det sena 80-talet*, Carlsson, Stockholm.

SKAF (1973), *En studie av medlemmar och organisation i kommunalarbetareförbundet*, SKAF, Report no. 1, Stockholm.

Söderpalm, Sven Anders (1975), 'The Crisis Agreement and the Social Democratic Road to Power' in Steven Koblic, ed.

Söderpalm, Sven Anders (1976), *Direktörsklubben. Storindustrin i svensk politik under 1930- och 40-talen*, Rabén & Sjögren, Stockholm.

Söderpalm, Sven Anders (1980), *Arbetsgivarna och Saltsjöbadspolitiken. En historisk studie i samarbetet på svensk arbetsmarknad*, SAF, Stockholm.

Söderqvist, Thomas (1986), *The Ecologists. From Merry Naturalists to Saviours of the Nation*, Almqvist & Wiksell International, Stockholm.

Sollerman, Erik (1958), 'Staten och trossamfunden' in FR.

SOU 1963, no. 17, *Författningsutredning. VI. Sveriges statskick. Del 2. Motiv. Förslag till regeringsform*, SOU, Stockholm.

SOU 1972, no. 15, *Nya regeringsform. Ny riksdagsordning*, SOU, Stockholm.

SOU 1990, no. 44, *Demokrati och makt i Sverige*, SOU, Stockholm.

SOU 1992, no 103, *Vinna eller försvinna– folkrörelsernas lotterier och spel i framtiden*, SOU, Stockholm.

Stolpe, Herman (1945), *Kooperationen och kvinnorna*, KFs bokförlag, Stockholm.

Stråth, Bo (1992), *Folkhemmet mot Europa. Ett historiskt perspektiv på 90-talet*, Tiden, Stockholm.

Stråth, Bo, ed. (1990), *Language and the Construction of Class Identities. The Struggle for Discursive Power in Social Organization: Scandinavia and Germany after 1800*, University of Gothenburgh, Department of History, Gothenburgh.

Ström, Lars-Inge (1985), *Föreningarna tar över. En studie av föreningsverksamhet som alternativ till kommunal verksamhet*, in Fi Ds 1985, no. 6.

Sundell, Jan-Olof (1966), *Unga Örnar*, Forum för Borgerlig Debatt, Stockholm.

Sundström, Erland (1952), *Trossamfund i det svenska samhället*, Tiden, Stockholm.

Svenska folkrörelser I. Nykterhetsrörelse, politisk arbetarrörelse, fackföreningsrörelse, folkbildning, kooperation (1936), Lindfors bokförlag, Stockholm.

Svenska Landskommunernas Förbund, Svenska Landstingsförbundet, and Svenska Stadsförbundet (1962), *Hundra år under kommunalförfatningarna 1862-1962*, Svenska Landkommunernas Förbund, Stockholm.

Svensson, Arne (1979), *De visade vägen. IOGT-NTO - 100 år. En krönika*, Sober Förlag, Stockholm.

Svensson, Bengt (1962), 'Socialinstitutets uppkomst' in Svenska Landskommunernas Förbund.

Swidler, Ann (1986), 'Culture in Action: Symbols and Strategies', *American Sociological Review* vol. 51: 273-86.

von Sydow, Björn (1989), *Vägen till enkammarriksdagen. Demokratisk författningspolitik i Sverige 1944-1968*, Tiden, Stockholm.

Tängerstad, Erik (1992/3), 'Hierarki och decentralering. Den svenska FNL-rörelsens förhistoria', *Arbetarhistoria* vol. 63-65: 28-34.

Tarrow, Sidney (1988), 'Old Movements in New Cycles of Protest: The Career of an Italian Religious Community' in Bert Klandermans, Hanspeter Kriesi, and Sidney Tarrow, eds.

Tarschys, Daniel (1978), *Den offentliga revolutionen*, Liber, Stockholm.

Therborn, Göran (1992), 'A Unique Chapter in the History of Democracy: The Social Democrats in Sweden' in Klaus Misgeld, Karl Molin, and Klas Åmark, eds.

Thomas, Clive S., ed. (1993), *First World Interest Groups: A Comparative Perspective*, Greenwood Press, New York.

Thulin, Henning (1945), *Pingströrelsen. En sociologisk orientering*, Förlaget Filadelfia, Stockholm.

Tilton, Tim (1992), 'The Role of Ideology in Social Democratic Politics' in Klaus Misgeld, Karl Molin, and Klas Åmark, eds.

Tingsten, Herbert (1967), *Den svenska socialdemokratins idéutveckling*, Aldus/Bonniers, Stockholm.

Tomason, Richard F. (1970), *Sweden: Prototype of Modern Society*, Random House, New York.

Torstendal, Rolf (1991), *Bureaucratisation in Northwestern Europe 1880-1985. Domination and Government*, Routledge, London.

Touraine, Allain (1986), 'Unionism as a Social Movement' in Seymour Martin Lipset, ed.

Trägårdh, Lars (1990), 'Varieties of Volkish Ideologies. Sweden and Germany 1848-1933' in Bo Stråth, ed.

UD (1992), *F 3-anslaget: 'Information, studier och forskning om fr4eds- och nedrustningssträvanden m.m.' En översyn*, Ministry of Foreign Affairs, Stockholm.

Uddhammar, Emil (1993), *Partierna och den stora staten. En analys av statsteorier och svensk politik under 1900-talet*, City University Press, Stockholm.

Vedung, Evert (1979), *Kärnkraften och regeringen Fälldins fall*, Rabén & Sjögren, Stockholm.

Vedung, Evert (1992), 'The Formation of Green Parties: Environmentalism, State, Response, and Political Entrepreneurship' in J. A. Hansen, ed.

Verba, Sidney et al. (1987), *Elites and the Idea of Equality. A Comparison of Japan, Sweden, and the United States*, Harvard University Press, Cambridge.

Vogel, David (1989), *Fluctuating Fortunes. The Political Power of Business in America*, Basic Books, New York.

Wahlström, Lydia (1944), 'Den kvinnliga rösträttsrörelsen' in Margareta von Konow, ed.

Wallin, Gunnar (1966), 'Folkomröstningsinstitutet' in *Samhälle och riksdag*.

Wallin, Gunnar (1991), 'Towards the Integrated and Fragmented State: The Mixed Role of Local Government' in Jan-Erik Lane, ed.

Weaver, Kent (1987), 'Political Foundations of Swedish Economic Policy' in Barry P. Bosworth and Alice M. Rivlin, eds.
Westerståhl, Jörgen (1945), *Svensk fackföreningsrörelsen. Organisationsproblem, verksamhetsformer, förhållande till staten*, Tiden, Stockholm.
Wikström, Tobias (1992), *De mindre, social politiska organisationerna*, Department of Political Science, University of Stockholm, Report for the project 'Swedish Interest Organizations', Stockholm.
Wittrock, Björn and Stefan Lindström (1983), *De stora programmens tid– forskning om energi i svensk politik*, Akademilitteratur, Stockholm.

For Product Safety Concerns and Information please contact our
EU representative GPSR@taylorandfrancis.com Taylor & Francis
Verlag GmbH, Kaufingerstraße 24, 80331 München, Germany